MATERIALS SCIENC]

RESEARCH ADVANCES IN MAGNETIC MATERIALS

MATERIALS SCIENCE AND TECHNOLOGIES

Additional books in this series can be found on Nova's website under the Series tab.

Additional e-books in this series can be found on Nova's website under the e-book tab.

MATERIALS SCIENCE AND TECHNOLOGIES

RESEARCH ADVANCES IN MAGNETIC MATERIALS

CARSON TOULSON
AND
DAMIEN MARWICK
EDITORS

New York

For permission to use material from this book please contact us:
Telephone 631-231-7269; Fax 631-231-8175
Web Site: http://www.novapublishers.com

Library of Congress Cataloging-in-Publication Data

ISBN: 978-1-62417-913-6

Published by Nova Science Publishers, Inc. † New York

CONTENTS

Preface		**vii**
Chapter 1	Principles of Creating Devices for Magneto-Laser Therapy with a High Magnetic Field Strength within the Optical Radiation Coverage Zone *V. Yu. Plavskii*	**1**
Chapter 2	Magnetic Nanostructures by Nano-Imprint Lithography *Saibal Roy and Shunpu Li*	**33**
Chapter 3	Photomagnetic Organic-Inorganic Hybrid Materials *Masashi Okubo and Norimichi Kojima*	**55**
Chapter 4	Spin-preference Rules for Conjugated Polyradicals *Masashi Hatanaka*	**77**
Index		**99**

PREFACE

In this book, the authors present current research in the study of magnetic materials. Topics discussed include the principles of creating devices for magneto-laser therapy with a high magnetic field strength within the optical radiation coverage zone; magnetic nanostructures by nano-imprint lithography; photomagnetic organic-inorganic hybrid materials; and spin-preference rules for conjugated polyradicals.

Chapter 1 – This paper covers methods of improving and fine adjustment of the magnetic induction of static magnetic field in the area of the optical radiation apparatus for magnetolaser therapy. It is shown that for the widely used in medical practice apparatus, that provide synergistic action of the given physical factors, the magnetic induction in the area of the radiation is 3,5–4,0 times less than its value over the projection of the body of the usually used ring magnets. The analysis of the distribution of the magnetic field on the surface of the magnetic nozzle apparatus for magnetolaser therapy using different types of cores of soft magnetic materials was carried out. Functional solutions and hardware implementation were proposed, that insure the formation in the area of the optical radiation of a magnetic field with induction, hardly different from its value over the projection of the body of the magnet. It is shown that there is a possibility of a smooth change in the range of the radiation magnetic induction in the range $|B|$ = 27–75 mT by to the change of the core relative to the plane of ring magnet. This adjustment allows to optimize the magnetolaser therapy taking into account the of pathology, age or individual physiological characteristics of patients. Principles of development of multifunctional equipment for low-intensity laser and magnetolaser therapy were developed for: local and zonal external magnetolaser therapy; internal laser therapy; photoblockage intrajoint and intratissue therapy; intravascular laser therapy using disposable intravenous tips; transcutaneous laser-induced blood therapy; laser-induced acupuncture of biologically active points. The apparatus is based on laser diodes and a collimator. Varying the collimator–diode distance regulates therapeutic efficacy. Internal therapy is based on an optical lightguide.

Chapter 2 – Patterned media, where the single domain recording bits are magnetically isolated from each other, offer the possibility to overcome many challenges faced by the other storage technologies, such as to obtain stable signal-to-noise ratio when the track width is reduced. The patterned media is likely to be the future media with high recording density and this has stimulated exploitation of different technologies to fabricate small featured magnetic materials. In parallel with the development of nanopatterning techniques there has been substantial progress in understanding and modelling the magnetic properties of small nanoparticles and patterned nano-dot arrays. The reduction of dimension leads to dramatic differences in micro/ nanostructure, coercivity, anisotropy, and magnetic moment etc which attracted enormous attention from both academia and industry in many areas and spans far beyond the information storage field. Hence the paper is organised in the following way: In section I, as the patterned media is the most promising application for nanomagnet array, the authors briefly overview the data storage principle and storage media. Section II describes several alternative imprinting and pattern transfer technologies. Section III introduces the spin configurations, magnetization reversal properties of small magnets. Finally, in section IV a summary of current and possible future development in this area is outlined.

Chapter 3 – The development of new magnetic materials is a key challenge in recent chemistry and physics. The design of magnetic coordination polymers has attracted much attention, because the intrinsic tunability of both the electronic and structural properties provides potential for multifunctional magnets. Therefore, the magnetic coordination polymers controllable by the external stimuli, which can be applied to electronic switching devices, have been investigated intensively. In particular, photomagnetism, *i.e.*, controllable magnetism by light irradiation, is one of the best studied multifunctional magnetisms. In this chapter, the recent progress in the development of photomagnetic organic-inorganic hybrid materials is summarized.

Chapter 4 – Since 1950's, prediction of spin-quantum number in conjugated radicals has attracted many theoretical chemists. There have been two major trends in spin-preference rules of conjugated biradicals. One is based on molecular orbital (MO) theory supported by Hund's rule. Another is based on valence bond (VB) theory, which has been intuitive for many chemists as a spin-polarization rule with spin alternation. Nowadays, as revealed experimentally and theoretically, localized orbital methods are better than the classical MO or VB methods in that some exceptional radicals that violate the simple MO or VB predictions are correctly described. Modern formalism of the spin-preference rule deals with exchange integrals between maximally localized MOs, which have been expanded from biradicals to general polyradicals by using Wannier functions. It has been shown that the best orbitals for description of the spin states of polyradicals are maximally localized Wannier functions, which minimize the exchange integrals. The Wannier analysis has been applied to many polyradicals and supported by semi-empirical, *ab initio*, and DFT calculations. In this chapter, construction and use of the Wannier-function method are reviewed.

In: Research Advances in Magnetic Materials ISBN: 978-1-62417-913-6
Editors: C. Toulson and D. Marwick

Chapter 1

PRINCIPLES OF CREATING DEVICES FOR MAGNETO-LASER THERAPY WITH A HIGH MAGNETIC FIELD STRENGTH WITHIN THE OPTICAL RADIATION COVERAGE ZONE

V. Yu. Plavskii[*]
B. I. Stepanov Institute of Physics,
National Academy of Sciences of Belarus, Minsk, Belarus

ABSTRACT

This paper covers methods of improving and fine adjustment of the magnetic induction of static magnetic field in the area of the optical radiation apparatus for magnetolaser therapy. It is shown that for the widely used in medical practice apparatus, that provide synergistic action of the given physical factors, the magnetic induction in the area of the radiation is 3,5–4,0 times less than its value over the projection of the body of the usually used ring magnets. The analysis of the distribution of the magnetic field on the surface of the magnetic nozzle apparatus for magnetolaser therapy using different types of cores of soft magnetic materials was carried out. Functional solutions and hardware implementation were proposed, that insure the formation in the area of

[*] E-mail: v.plavskii@ifanbel.bas-net.by.

the optical radiation of a magnetic field with induction, hardly different from its value over the projection of the body of the magnet. It is shown that there is a possibility of a smooth change in the range of the radiation magnetic induction in the range $|B|$ = 27–75 mT by to the change of the core relative to the plane of ring magnet. This adjustment allows to optimize the magnetolaser therapy taking into account the of pathology, age or individual physiological characteristics of patients. Principles of development of multifunctional equipment for low-intensity laser and magnetolaser therapy were developed for: local and zonal external magnetolaser therapy; internal laser therapy; photoblockage intrajoint and intratissue therapy; intravascular laser therapy using disposable intravenous tips; transcutaneous laser-induced blood therapy; laser-induced acupuncture of biologically active points. The apparatus is based on laser diodes and a collimator. Varying the collimator–diode distance regulates therapeutic efficacy. Internal therapy is based on an optical lightguide.

1. The Current Status of Low Level Laser- and Magneto-Laser Therapy

1.1. The Place and Role of Low-Level Laser Therapy in the Arsenal of Resources of Modern Medicine

The first data that demonstrated the biological action and therapeutic efficacy of low-level laser radiation (LLLR) appeared in the late 1960s and early 70s [1–10]. They were obtained using as a radiation source a helium-neon laser λ = 632.8 nm – at that time one of the few commercially available lasers that were relatively cheap, reliable in operation, and easy to control. For the next thirty-five years, laser therapy was extensively developed and occupied a special place among the therapeutic physical factors so far used in medical practice. The range of use of low-intensity lasers in medicine (treatment, rehabilitation, prophylaxis) is currently so great and the spectrum of therapeutic action of the indicated physical factor is so wide that it has become taken for granted to speak of the creation of a new major specialization of modern physical therapy – low-level laser therapy (LLLT) [11–19]. According to the concepts that have become stabilized in the medical and biophysical literature, low intensity is taken to mean radiation that corresponds to the range of intensities that show therapeutic action on the organism not because of thermal processes. As a rule, the mean power of

LLLR does not exceed 500 mW. Here there is a certain disagreement in the literature in questions concerning the spectral range of radiation that corresponds to the category of LLLR. Thus, in English-language sources, this category includes only radiation in the red and near-IR regions (wavelength from λ =632.8 nm of the helium-neon laser to λ = 1064 nm of the Nd:YAG laser). At the same time, radiation in the UV region (for example, λ = 308 nm of an excimer laser), which is fairly widely used for the treatment of skin diseases, is not included in the category of LLLR. However, in the Russian literature, along with red and IR radiation, the indicated physical factor also includes long-wavelength UV radiation (for example, λ = 325 nm and λ =337.1 nm from the helium-cadmium and nitrogen lasers, respectively) and radiation of the blue-green region (for example, λ = 441.6 nm, λ = 510.6 and 578.2 nm of the helium-cadmium laser and the copper-vapor laser, respectively).

It should be pointed out that the list of diseases where the use of LLLR has a beneficial character is continuously expanding as data are accumulated concerning the mechanisms that determine the therapeutic action of laser radiation and as laser engineering and fiber-optic methods for delivering it to the foci of injury at various localizations undergo development and improvement. Clinical studies of recent years have demonstrated the successful use of the indicated physical factor in the treatment of pathologies (oncological diseases [20–22], tuberculoses [23, 24], the effects of sugar diabetes [25], perinatal illnesses of the newborn [26] etc.) that were considered counterindications to its use 10–15 years ago. The factors that increase the popularity of LLLT among patients and cause it to be used everywhere are (a) its high therapeutic effectiveness and the wide spectrum of indications to the application of the method, including the treatment of chronic and degenerative-dystrophic diseases, when drug therapy alone is insufficiently effective; (b) the absence (in the overwhelming majority of cases) of the side effects inherent to many pharmaceutical preparations; (c) the possibility of using LLLT in combination with drug therapy and other physical-therapy factors; (d) the presence of pronounced pain-relieving (anesthetizing) action; (e) the possibility of carrying out a procedure of laser therapy (intravenous or supravenous transcutaneous action on the blood; action on biologically active points and zones, etc.) when the focus of injury is not accessible to the direct action of laser radiation (because of the presence of bandages or plaster casts); (f) the noninvasive nature and comfort of most laser therapeutic procedures; (g) availability of the apparatus.

The fact that LLLT is a clinical reality and not a myth is confirmed by the positive results of the treatment of patients with various pathologies, described by many authors of more than 3000 publications (see, for example, Refs. [11–26]. Also, as shown by an analysis using the databases Medline, Embase, Cochrane Library, PubMed, Biological Abstract, and BIOSIS of more than 140 clinical studies (see, for example, Refs. [27–37]), conclusions concerning the effectiveness of LLLT methods are obtained from the viewpoint of demonstrative medicine (randomized clinical testing), while some of them are obtained with so-called double-blind control, when neither the physician nor the patient is allowed to know which of the versions uses a placebo and which uses laser action.

However, it should be pointed out that, despite copious data on the successful use of laser radiation to treat diseases of various etiologies, there are also indications in the literature, including some obtained from the viewpoint of demonstrative medicine (randomized studies [38–41]) demonstrating that the action of laser radiation with respect to therapeutic effectiveness is virtually no different from the placebo effect (i.e., simulation of laser radiation). In our opinion, this fact is evidence, first, that laser radiation, like any other physical factor or pharmaceutical method, is not a panacea and, second, that, to obtain pronounced therapeutic action of LLLR, it is necessary to follow definite regimes and operating techniques [42]. Tunér and Hode in their review [43], which critically analyzes the possible causes of the absence of therapeutic action of LLLR described in a number of literature sources, attribute the absence of a reliable therapeutic effect to the fact that the parameters of the active radiation are far from optimal.

For a long time, the main “argument” of skeptics of the method of LLLT was the assertion that it was not very “popular” in the leading western countries. Actually, until 2002 LLLT was widely used only in the USSR (the CIS), China, Italy, Australia, Hungary, and some other countries, while its use had a restricted (research) character in the USA, because it had not been approved by the US Food and Drug Administration (FDA). In the FDA’s opinion, the factor that militated against a positive decision on the use of LLLR was inadequate scientific understanding of the mechanism of the therapeutic action of the indicated physical factor. In this case, according to the repeated statements of the FDA, the presence of copious data from various countries on positive results of the use of LLLR for therapeutic purposes cannot serve as a basis for issuing a decision on its use in the USA. Nevertheless, one of the first laser therapeutic devices to receive precommercial notification of the FDA in January 2002 was the Microlight

830 Laser System, which had been available for years in Europe from the Danish firm CB Medico and was presented to the FDA for certification by the Micro-Light Corporation of America. As shown by a retrospective analysis, the decision by the FDA that LLLR methods could be used in US medical facilities gave a new impulse to the activation of developments of laser devices in other countries, including those on the European continent.

1.2. Factors Determining the Effectiveness of the Action of Laser Radiation and Static Magnetic Field

Magneto-laser therapy is a combined effect on the body with curative purposes by magnetic field and low-intensity laser radiation. The method was proposed by A.K. Polonskii and co-authors in 1977 [44-46]. As it is known, combined physiotherapeutic methods should be based primarily on the biological effects of the synergies of the combined in a single procedure of therapeutic physical factors. Both magnetic field and laser radiation can be characterized by anti-inflammatory, analgesic, anti-edema, immune and other effects. This similarity of therapeutic effects assumes their intensification (synergy) with the simultaneous use of these physical factors. In addition, both physical factors affect in one direction the number of metabolic and physiological processes: microcirculation, blood rheology, blood permeability, activity of the endocrine organs, synthesis of high-energy phosphates, exchange of proteins, nucleic acids, etc. With the combined use of low-intensity laser radiation and a static magnetic field along with summing up of unidirectional physiological and therapeutic effects, a number of physical, chemical and biophysical changes arise, that are important to base magnetic laser therapy and understanding of the kind of effects on the body [47-49].

Taken place under the action of magneto-laser therapy primary physical and chemical changes are accompanied with biological reactions that affect various body systems. The most important and proved are: activation of biosynthetic processes and formation of energy-rich phosphates, changes in vascular permeability, microcirculation and peripheral circulation, degranulation of mast cells, formation of physiologically active substances and changes in hormonal balance, stimulation by the means of direct influence and reflex mechanism of functional mode of organs and systems and others [47-49].

The main therapeutic effects of magneto-laser therapy: anti-inflammatory, analgesic, immunotherapy, antispastic and antihypoxic. Decrease in cholesterol, higher levels of antioxidants, increase of prostaglandin synthesis, decreased lipid peroxidation take place by magnetic-therapy. All above also largely determines its therapeutic effect.

To implement this method different variants of combination of magnetic fields and laser light are used [15, 42]. Most often static magnetic field with continuous or pulsed low-intensity laser light of red and near-infrared regions of the spectrum is combined. There are also devices for combined action of laser radiation and alternating magnetic field, or devices that provide exposure to laser radiation with the static and variable magnetic fields [42].

Technique and methods of magnetic laser therapy are similar to laser therapy and consists in exposure on the area of pathological focus, on cutaneous projection of bodies, on reflex zones and acupuncture points, as well as intracavitary techniques. It can also be used for above-vessels blood irradiation. Exposure is most often taken place by the means of a stable (stationary) method, contactly, using one or more fields. In the treatment of wounds, venous ulcers, burns and skin diseases non-contact exposure is used. In this case, the transmitter is to be located at a minimum distance from the irradiated surface [47-49].

Magneto-laser therapy is most often used at laser power density of 50-100 mW/cm^2 and magnetic induction in the range of 20-50 mT. Exposure of one field is usually about 3-5 minutes, while the total duration is not more than 15-20 minutes. The treatment course consists of 8-12, rarely 15-18 procedures performed daily or every other day [47-49].

The list of diseases treated by laser and magnetic therapy is very broad and growing constantly. Most successfully listed physiotherapeutic procedures are used for the following diseases: in surgery – nonhealing wounds, sores, burns, frostbite, vascular diseases of the lower extremities; in trauma and orthopedics – inflammatory and traumatic diseases of the joints and spine, bone fractures, myalgia, arthralgia; in dentistry – gingivitis, stomatitis, pulpitis, periodontitis; in therapy – coronary heart disease, hypertension, non-specific inflammatory respiratory diseases, peptic ulcer, chronic gastritis, hepatitis, colitis; in neurology – neuralgia, neuropathy, spinal osteochondrosis with neurologic manifestations; in obstetrics and gynecology – lactational mastitis, infertility, inflammatory diseases of the internal organs, and in dermatology – allergic dermatitis, atopic dermatitis, psoriasis, eczema, lichen planus, recurrent herpes, acne vulgaris, etc [15, 42, 44-59].

Of course, laser therapy and magneto-laser therapy is not an alternative to medical and surgical treatment. However, the inclusion of this method into the combined treatment not only allows to achieve a more pronounced and more rapid positive effect, but also reduces the frequency and severity of side effects of medicines used by decreasing the required dose.

2. Correction of Magnetic Field Distribution within the Optical Radiation Coverage Zone of Magnetic Laser Therapy Apparatuses

As it has been already noted, the synergy effect of low-intensity laser radiation and constant magnetic field is observed only in the case of simultaneous (joint) use [15, 42, 44 - 49]. Consistent exposure to laser radiation and magnetic field does not cause a noticeable increase in regulatory activity of effectors.

It is considered [42, 60-67], that synergy is underlied by equivalence (generality) in the mechanisms of the biological action of physical factors, which consists in the orientation effect of the electric field of the laser light wave and magnetic field on biological structures with liquid crystal character ordering (the cell membrane, multienzyme complexes), responsible for metabolism regulation. Meanwhile, sensitivity of supramolecular formations to light-induced conformational changes increases in the presence of magnetic field [42, 60 -67]. Therefore, to achieve a pronounced therapeutic effect with the use of devices for magneto-laser therapy, exposure area to laser radiation should correspond to the area of magnetic field application.

There is a number of facts [42, 60-62], indicating that the increase of magnetic field gives the same biological and therapeutic effects at a lower intensity of optical range. It ensures a more pronounced light-induced regulatory effect on lesion focuses of deep localization, delivery of high-intensity optical radiation to which is problematic due to light scattering and absorption by proper tissues and skin.

In medical practice magneto-laser therapy method is realized, as a rule, through the use of magnetic nozzles of a ring (circular) form on laser devices [15, 42]. To affect the pathological focus laser beam passes through the cavity in a ring magnet (usually through its axis of symmetry). As analysis of the data references [15, 67-71], by the given constructive solution of devices for magneto-laser therapy, intensity of magnetic field at the surface of the ring magnet is minimal in the range of optical radiation, and its maximum value is recorded on the body surface of the magnet.

Furthermore, the only practical way to adjust the intensity of magnetic field in the area of optical radiation is the change in the distance of the magnet to pathological focus. In addition, for devices with magnetic-uncollimated (divergent) laser beam distance change affects not only the strength of magnetic field, but also the power density (intensity) of optical radiation, which significantly complicates to control parameters of influencing factors during the magneto-laser therapy.

The purpose of this part of the article is to work out principles of making a device for magneto-laser therapy with higher magnetic induction in the area of optical radiation and with possibilities of its fine adjustment over a wide range.

2.1. Material and Methods of Measurement

Magnetic induction distribution was tested in a strontium ferrite tip to ring magnet K36×18×6 (Republican United Company Ferrit, Minsk). The outer diameter of the magnet is 36 mm. The inner diameter of the magnet is 18 mm. The height of the magnet is 6 mm. The tip is made of D16 alloy. Magnetic induction was monitored using a TP2-2U portable milliteslameter. The sensor was attached to the tip of the magnet.

The experimental setup is shown in Figure 1. The sensor moves from tip center to tip periphery along radius r (Figure 2). The gap between magnet 2 and sensor 9 was $d = 0.5 \pm 0.02$ mm (Figure 2). Experimental error of the TP2-2U milliteslameter was ±2.9%.

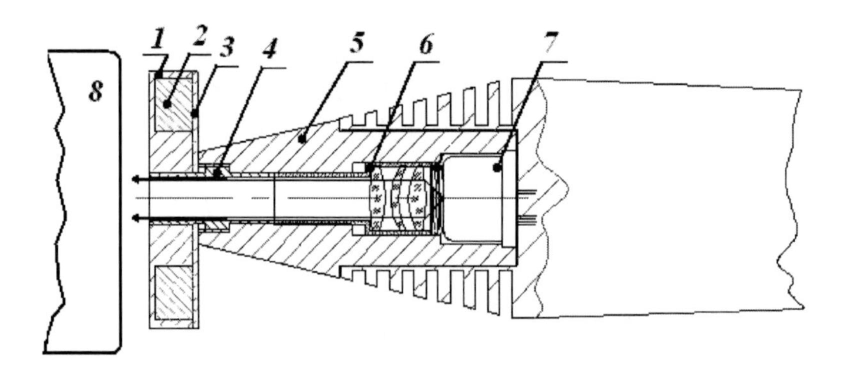

Figure 1. Diagram of arrangement of laser and magnetic tip in magnetotherapeutic apparatus: 1) tip body; 2) ring magnet; 3) magnetic soft core made as a plate with holes directed to a laser diode; 4) axial tubular core; 5) laser source body; 6) laser collimator; 7) laser diode; 8) object.

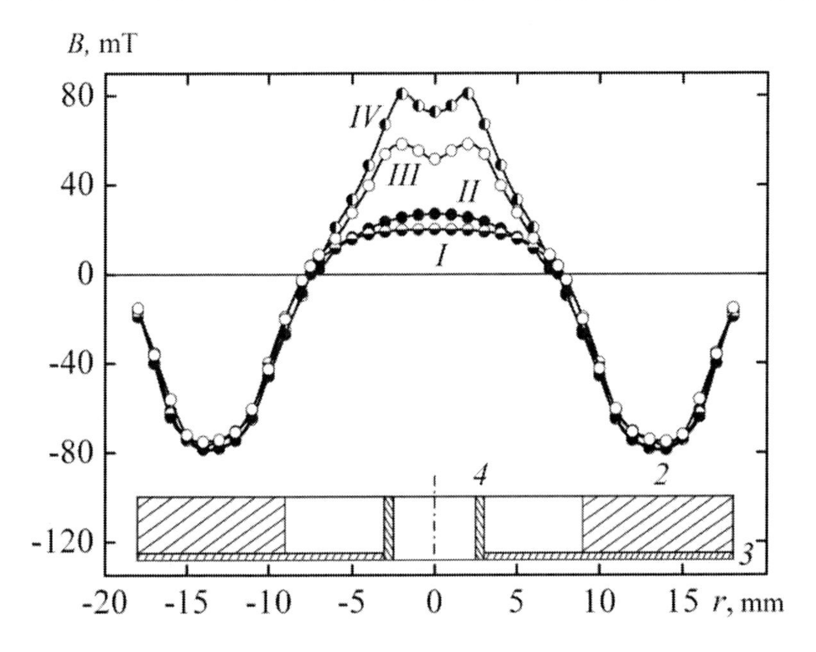

Figure 2. Diagram of magnetic induction distribution at tip surface: I) without core; II) plate core 3 with hole directed to laser diode; III) axial tubular core 4; IV) plate core 3 with hole directed to laser diode and axial tube 4; 9) sensor of milliteslameter.

2.2. The Distribution of the Static Magnetic Field in the Apparatus for Magneto-Laser Therapy and Methods of Its Correction

Magnetic induction B distribution over the surface of tip 1 (Figure 2, variant I) is shown in Figure 3, curve I. Tip surface point directed to the object is taken as the zero point along axis r (Figure 3). The value of r increases upon moving the milliteslameter sensor from the tip center to the periphery. Curve I in Figure 3. shows magnetic field distribution (tip 1 surface). The magnetic field is generated byring magnet 2. Movement of the milliteslameter sensor 9 from the tip center to periphery changes magnetic field polarity and magnetic induction sign. Induction B is positive or negative at the tip center or at ring magnet projection 2. Magnetic induction is maximal at distance r = 12-15 mm (Figure 3, curve I). This maximal value is $|B|$ = 72-77 mT, which is observed at the projection of magnet 2. Magnetic induction at the magnet center is $|B|$ = 20 mT. In the absence of the core, magnetic induction in the center of the ring magnet is 3.5-4-fold lower than at the projection of magnet 2.

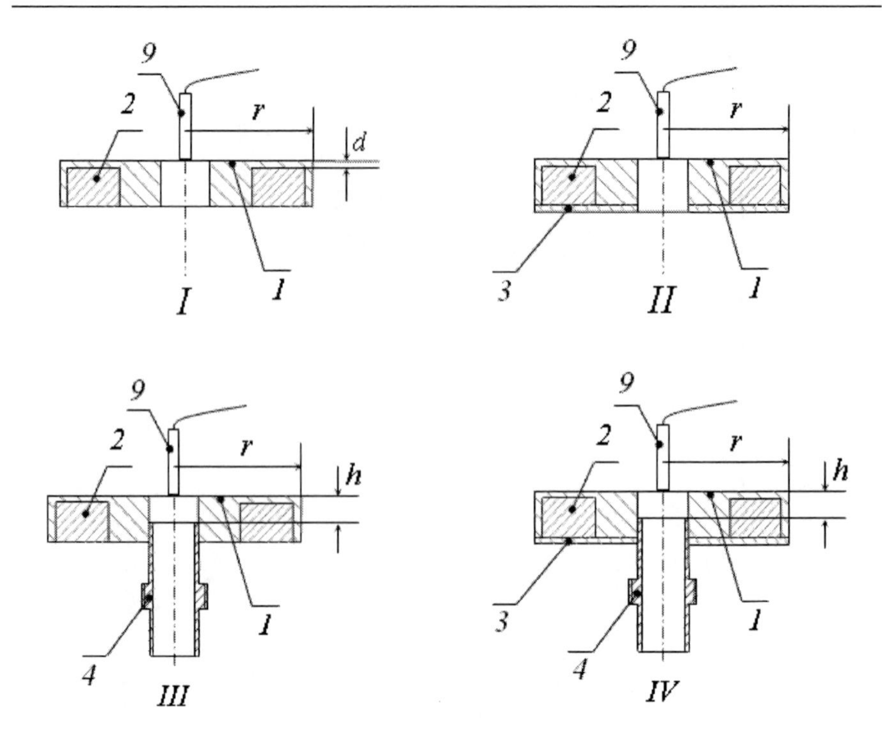

Figure 3. Magnetic induction distribution over tip surface relative to magnet 2 projection and tubular core 4 for different structures: I) without core; II) with core 3 as plate with holes directed to the laser diode; III) with axial tubular core 4; IV) with core 3 as a plate with holes directed to the laser diode and axial tube 4.

Variant II (Figure 2) of tip 1 involves plate 3 made of 21880 steel with a hole directed to the laser. The plate is located at magnet 2. This plate modifies magnetic induction distribution over tip 1. This result is shown in Figure 3. (curve II). Magnetic induction at the center of the tip 1 is $|B| = 27$ mT. Magnetic induction at the magnet projection is $|B|$ = 74-78 mT. Plate 3 increases magnetic induction at the center of tip 1. This value is 3-fold lower than the magnetic induction at the projection of magnet 2.

The 21880 steel tubular core 4 causes more pronounced changes in magnetic induction distribution over the surface of tip 1. This core is located along the axis of magnet 2 (Figure 2, variant III). Magnetic induction distribution with tubular core 4 ($h = 0$) is shown in Figure 3 (curve III). Plate 3 was not attached to tip 1. Tubular core 4 increases magnetic induction at the center of tip 1 to $|B| = 52$ mT. At distance $r = 2.5$-3.0 mm from the center of tip 1 (at the projection of tube 4) the magnetic induction is even larger, $|B| = 58$

mT. The magnetic induction above the projection of magnet 2 was constant ($|B|$ = 70-75 mT) at r = 12-15 mm.

A double core (tube 4 + plate 3) provides maximal magnetic induction at the laser beam (center of magnetic tip 1). This arrangement is shown in Figure 3. (variant IV). Magnetic induction distribution over the surface of tip 1 at h = 0 is shown in Figure 3. (curve IV). With the combined core magnetic induction at the center of tip 1 is 73 mT. This value is close to $|B|$ = 74-78 mT at the surface of magnet 2 at r = 12-15 mm. At tube 4 (r = 2.5-3 mm from center of tip 1) $|B|$ = 78 mT.

The results shown in Figure 3. indicate that maximal magnetic induction is provided by a magnetic core including axial tube 4 located on the axis of ring magnet 2 and plate 3 with a hole at magnet 2 directed to the laser.

The magnet tip provides graduated regulation of magnetic induction without changing tip–object distance. Regulation of magnetic induction can also be achieved by variation of the distance h of tip 1 to tube 4. Dependence of magnetic induction in the center of tip 1 at r = 0 for magnet–tube distance h for variants III and IV (Figure 2) is shown in Figure 4. When the tip surface is located near the tube, h = 0. Curves III and IV in Figure 4 show magnetic induction dependence on h without and with plate 3.

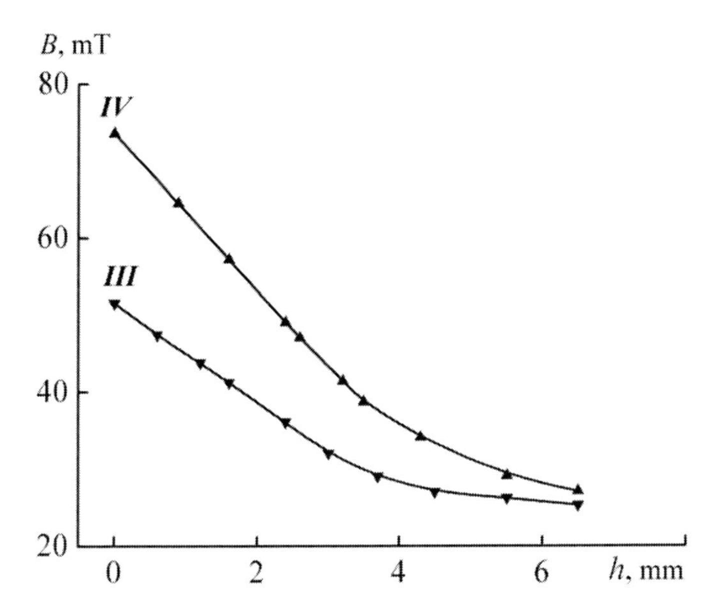

Figure 4. Dependence of magnetic induction at the center of the tip on magnet–tube distance h: III) axial core made as tube 4; IV) core made as plate 3 with hole directed to the laser diode and axial tube 4.

In the absence of plate 3 the magnetic induction at the center of tip 1 is lower than in the presence of the plate. For the combined core (Figure 2, variant IV) containing plate 3 and tube 4 the magnetic induction range is 27-74 mT (Figure 4, curve IV). In the absence of plate 3 (Figure 2, variant III) h value provides magnetic induction in the range 25-52 mT (Figure 4, curve III).

2.3. Conclusion

We have investigated the distribution of constant magnetic field on the surface of the magnetic nozzle (including the area of the optical radiation) in the apparatus for magneto-laser therapy, and developed methods for correcting the magnetic induction. It is shown that for the widely used in medical practice apparatus, having an expressed therapeutic effect because of the synergy of these physical factors, the magnetic induction in the area of the optical radiation is 3,5-4,0 times less than its value over the projection of the body of the commonly used in such installations ring magnets. To increase the intensity of the magnetic field in the area of the laser radiation it is proposed to use various types of hollow cores of soft magnetic materials, capable to have a significant impact on the character of the magnetic field distribution on the surface of the nozzle. Constructive solutions of the magnetic nozzle with combined cores were proposed to provide the formation (in the area of laser radiation) of the magnetic field with induction, slightly different little from its value over the projection of the body of the magnet. These magnetic nozzles can also be used for magneto-phototherapy in devices based on LED sources. It's shown, that there is the possibility of a smooth change (in the area of the radiation) of the value of magnetic induction in the range $|B| = 27\text{-}75$ mT due to the change of the core in regard to the plane of ring magnet. This adjustment allows to optimize the magneto-laser therapy taking into account pathology, age or physiological characteristics of individual patients.

3. Principles of Development of Multifunctional Equipment for Low-Intensity Laser and Magnetolaser Therapy

Modern technologies of laser and magnet-laser therapy involve various methods of exposure to optical radiation to achieve its regulatory effect on the human body: exposure on lesions of external localization, exposure on the projection of internal organs through skin, exposure on lesions of intracavitary localization; abovevenous (transcutaneous) exposure on blood; intravascular (intravenous) blood irradiation using a disposable sterile fiber optic systems, intra-and interstitial (photoblockage intrajoint and intratissue therapy) effects, exposure on biologically active points (laser acupuncture) and reflex zones (Zakharin-Head zones) [11-18].

Laser diodes (semiconductor lasers) of visible and near-infrared regions of spectrum are mainly used as sources of radiation in phototherapy devices. These semiconductor emitters practically superseded other types of lasers from modern mass-produced (certified) therapeutic devices thanks to the possibility of selecting a wavelength in a wide range, compactness, lack of high-voltage power supply, ease of implementing the devices that does not require grounding, low power consumption (making it possible to work on internal independent power supply - small batteries), absence of fragile glass components, everpresent in gas lasers; easily realizable possibility to change affecting parameters (output power, pulse frequency, or frequency modulation), reliability and durability, relatively low cost and commercial availability [42].

However, the use of laser diodes as a light source of multifunctional therapeutic devices creates certain difficulties for carrying out the designs that provide the possibility of exposure by using a single radiator both on lesions of external and intracavitary (interstitial) localization, as well as for intracorporeal exposure of blood. The reason of emerging problems is a high divergence in laser diodes, that is different in the plane of p-n-junction and in the perpendicular plane.

The simplest approach, that was widely used in the first therapeutic devices based on semiconductor lasers, was to use for the delivery of radiation to the patient of a long (~ 1.0-1.5 m) monofilament fiber (with a diameter of ~400-600 μm light-guiding core) strictly fixed to the radiator, the distal end of which is placed in the handle crane. The disadvantage of this system is almost complete depolarization at the distal end of the fiber [42], that according to [61- 64, 72-80] can significantly affect the biological and therapeutic effects of light.

For this reason, phototherapeutic devices have been developed and widely applied that provide exposure to laser radiation of pathological lesions of cavernous localization via a short replaceable fiber optic tools, optically bounded with a diode emitter [12, 15, 42]. In this case (in contrast to the first generation of therapeutic devices based on semiconductor lasers) diode emitter is spatially separated from the power supply and connected to it by a flexible electrical cable. Typically, the light guide attachment in this case is made of straight or curved stainless steel tube with a length 11-18 cm, inside of which a quartz-polymer monofiber with a diameter of 0.6-1.0 mm is located. Most often, the input of laser diode into the fiber is carried out without the use of additional optical elements (thanks to bringing of the input end of the optical fiber directly to the output window of the radiation source), sometimes – by means of a short- focus lens, focusing on the input end of light guide attachment [12, 15, 42]. This technical solution is advantageous because of slightly reduced degree of polarization at the fiber output. It is $p \approx 0.9$ (while at the laser diode output $p \approx 0.8$) [42]. External exposure in this case is carried out without any fiber optic tools by manipulating the actual laser emitter. The disadvantage of this design solution is the need to maintain a fixed distance between the laser source and the surface of the body during the entire process of phototherapy sessions (lasting 5-10 minutes). Otherwise, due to strong divergence of radiation (which is up to 45° that in laser diodes) change in the distance of emitter - surface of the body leads to a change in the area of light spot, and accordingly – in the change of power density of affecting radiation, which determines effectiveness of treatment. In addition, the use of highly divergent radiation does not allow an effective exposure on the deeply localized lesions due to a rapid decrease in the intensity of the influencing factor in the tissue. In this regard, a number of authors [11, 42] recommends to carry out external non-invasive impacts not by laser radiation of low power density out-of-focused over the entire pathological lesion, but by a collimated beam, providing a relatively high intensity of affecting radiation, independent on the distance of the emitter-surface of the body. In this case, the effect takes place at several points over the pathological lesion with optimal power density flow. However, the known devices based on semiconductor lasers as a rule do not allow to use collimated light. The analysis of known structural approaches to the construction of laser therapeutic equipment shows that implementation of the scope of methods of exposure to laser radiation for phototherapy purposes using a single radiator is often either ineffective or requires the use of a set of small removable optical elements (microscope objectives with different focal lengths), which creates certain discomfort when using the product. The objective of this work is working out the principles of designing a multifunctional multicolor phototherapeutic device based on laser diodes, which provides opportunity for effective laser and magnetolaser therapy, regardless of the localization of patient's pathological lesion, and eliminates the drawbacks of known devices based on semiconductor lasers.

3.1. Specificity of Laser Radiation Source with Mobile Collimator for Therapy

The mobile collimator is a specific element of the laser and magnetotherapeutic system developed in this work. The divergence of the laser radiation is compensated by the collimator [69]. The phototherapeutic apparatus (Figure 5, a-c) provides phototherapy in the presence of a magnetic field external to the pathological foci.

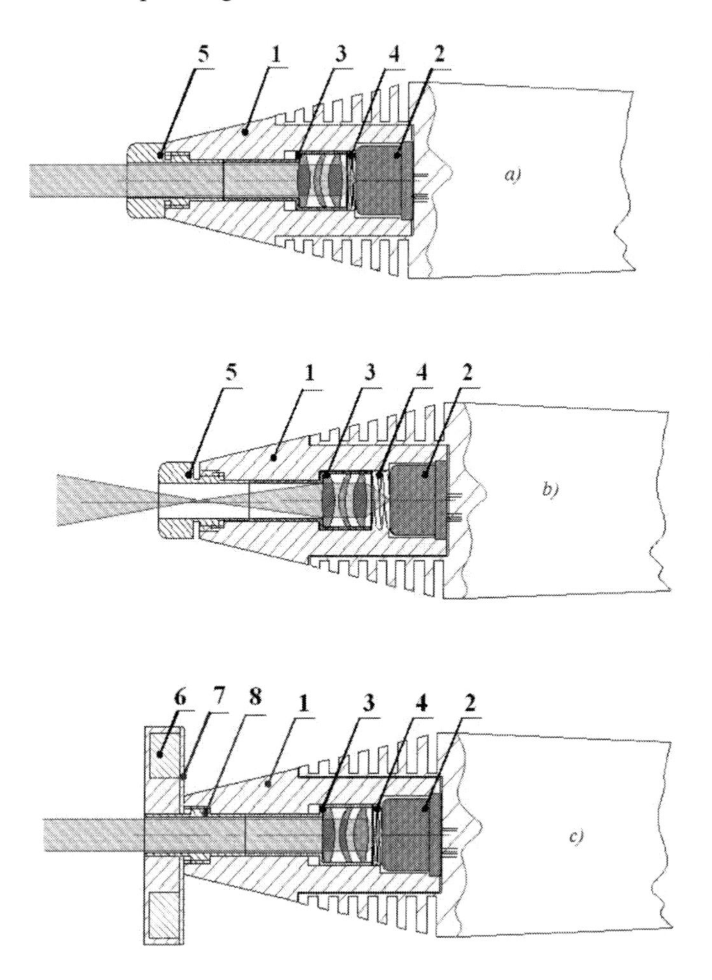

Figure 5. Radiation source of laser phototherapeutic apparatus for external pathological foci using collimated (a, c) and divergent (b) radiation with background of magnetic field (c): 1) radiation source body; 2) laser diode; 3) optical collimator; 4) spring; 5) focusing tip; 6) magnet tip.

Figure 5. shows that the apparatus contains radiation source 1 and laser diode 2, as well as collimator 3, which can be moved along the optical axis. Spring 4 is placed between laser diode 2 and collimator 3. The spring does not block the propagation of the radiation. Focusing tip 5 is at the laser diode body 1. The tip is regulated using a screw.

Spring 4 regulates the distance of laser diode 2 from collimator 3. Tip 5 determines this distance. The diameter of collimator 3 is adapted to the hole diameter in body 1.

The collimator 3 is moved with spring 4. Parameters of collimator 3 and laser diode 2 are adapted to the aperture of the collimator lens (Figure 5.b). Focusing tip (Figure 5, a and b), magnet tip (Figure 5.c), acupuncture tips (Figure 6), and lightguide (Figure 7) provide phototherapy.

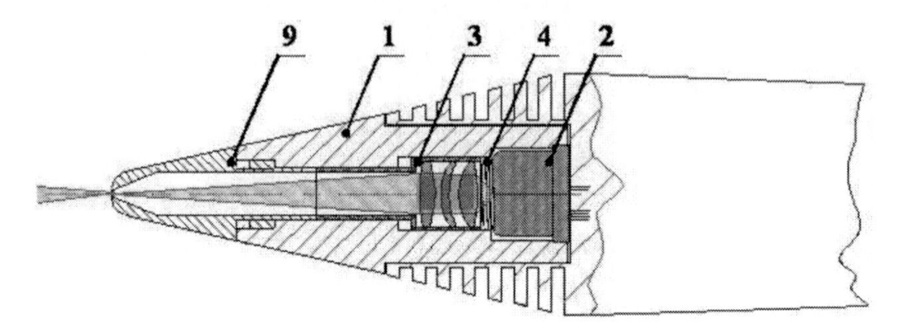

Figure 6. Laser radiation source of phototherapeutic apparatus for acupuncture: 1) radiation source body; 2) laser diode; 3) optical collimator; 4) spring; 9) acupuncture tip.

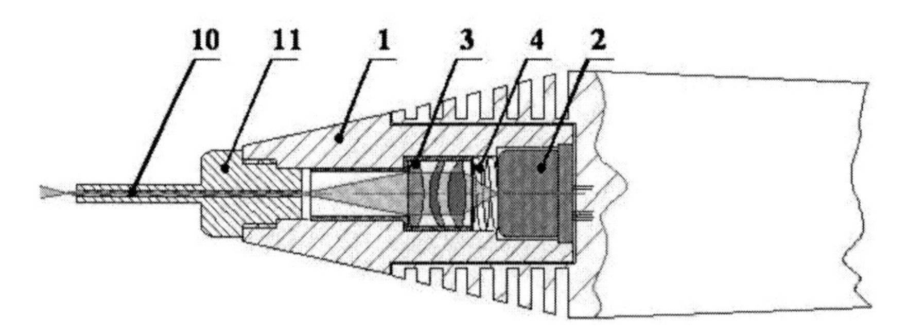

Figure 7. Laser radiation source of phototherapeutic apparatus for internal and intravascular phototherapy using monofiber lightguide: 1) radiation source body; 2) laser diode; 3) optical collimator; 4) spring; 10) monofiber lightguide; 11) tip for monofiber lightguide.

Figure 5.a, b. shows that the focusing tip 5 regulates radiation divergence. Collimator 3 moves toward laser diode 2 using spring 4. The screw of tip 5 provides minimal radiation divergence (collimated beam) (Figure 5.a). In this case, laser radiation density does not depend on laser–patient distance. Spring 4 varies laser diode–collimator distance. The diameter of the light in the patient's body is also varied (Figure 5.b).

The mobile collimator allows the light spot diameter to be varied without changing the patient–light source distance.

Focussing tip 5 can be incorporated in magnet tip 6 (Figure 5.c). The magnet tip also changes the collimator–laser diode distance. Activity of magnetic tip 6 with cores 7 and 8 was reported in [69]. A magnet tip with induction B = 75 mT was described in [70]. Combination of laser -radiation with magnetic field is synergistic [42, 44-49, 61, 62, 64, 70]. The effect of synergism is due to orientation effect of laser radiation on macromolecules and membranes. Magnetic field increases sensitivity of conformational dynamics of macromolecules. This effect was reported in [61, 62] using mitotic cultures exposed to: a) laser radiation; b) static magnetic field; c) combination of laser radiation with magnetic field. The combination of laser radiation with magnetic field induces more pronounced effect than the individual factors. Optimal therapeutic effect in the presence of magnetic field is observed at lower laser radiation intensity than in the absence of magnetic field. Such synergism is also observed in fish embryos [64].

Transcutaneous laser magnetotherapy is widely used in clinics [11, 14, 12-18, 44-49, 81]. The apparatus described in this work contains ~5 mm tip for phototherapy. This tip is attached to the patient's elbow. A circular magnet is inserted into the tip. The laser is fixed during the therapy. Radiation density for subcutaneous therapy is 80-100 mW/cm^2. Red radiation of 670 nm is collimated. This radiation is superimposed with magnetic field 50-75 mT. This technology is used in therapy of cardiological patients (arterial hypertension, ischemic heart disease, myocardial infarction, acute coronary insufficiency, cardiac failures, etc.).

Tip 9 instead of tip 5 is used for acupuncture therapy (Figure 5, a and b; Figure 6). The size of tip 9 is adapted to the size of the collimator and laser diode. The acupuncture tip is superior to monofiber tips. Long-term use of the acupuncture tip requires its cleaning.

The apparatus described in this work provides internal and intravascular laserinduced therapy (Figure 7). Monofiber lightguide 10 is in tip 11. The tip provides radiation of light source 1 collimation (using collimator 3) and focusing of laser diode 2 onto monofiber lightguide 10. In the case of intracavitary use (oral cavity, nasal cavity, etc.), the lightguide is 10-15 cm long and accommodated in a stainless tube (diameter $d \approx 1.0$ mm). Blood is exposed intravenously using a 18-20-cm sterile disposable lightguide [82]. This lightguide is attached to tip 11. The lightguide diameter is 100-150 μm (internal needle diameter is $d \approx 700$ μm). Phototherapy of gastric ulcer, duodenal ulcer, etc. is also possible. A lightguide with diameter $d \approx 400$ μm is inserted into a polymer catheter (diameter $d \approx 1.6$ mm).

Tip replacement does not require additional adjustment.

3.2. Features of Laser Module Providing Variation of Radiation Wavelength

Laser therapy, acupuncture, and intravenous exposure of blood are mediated by radiation of semiconductor lasers of various wavelengths (Figure 8, a and b).

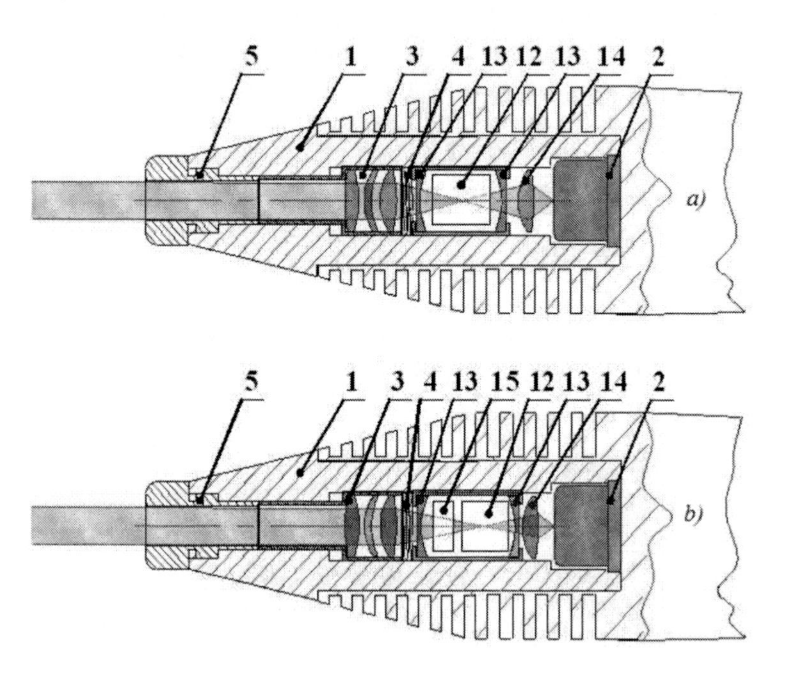

Figure 8. Laser radiation source of phototherapeutic apparatus for external exposure within different spectral ranges: 1) radiation source body; 2) laser diode; 3) optical collimator; 4) spring; 5) focusing tip; 12) solid-state active elements; 13) mirrors of resonator; 14) focusing lens; 15) nonlinear crystal.

Radiation of laser diode 2 is transformed in active element 12 in reflector cavity 13. Focusing tip 14 decreases light spot diameter to a few millimeters. Spectral range is determined by the laser diode 2 and solid-state element 12. Rare-earth-ion-doped crystals (Nd^{3+} and $Tm^{3+)}$ are used as active elements for the AlGaAs laser diode (λ = 805-808 nm). Laser generation at 0.94, 1.06, 1.32 μm requires active element yttriumaluminum garnet doped with Nd. Laser generation at 1.9 μm requires active element K-yttrium-tungstate doped with Tm [83-85]. Figure 8.a. shows that generation wavelength can be longer than pumping wavelength. It follows from Figure 8.b. that laser diode 2 wavelength can be transformed to shorter wavelength in cavity 13 using active element 12 and crystal 15. Optical harmonics generated by a nonlinear crystal convert IR laser radiation into visible light. A laser diode with λ = 805-808 nm based on yttrium-aluminum garnet doped with Nd using an LBO crystal generates at 0.47 μm (second harmonic of 0.94 μm), at 0.53 μm (second harmonic of 1.06 μm), or at 0.67 μm (second harmonic of 1.3 μm) [83, 84]. Variation of laser radiation wavelength increases therapeutic efficacy [1, 11-18]. The spectral range 600-1200 nm is a biological transparency window [14, 42]. The laser radiation penetration depth within this window is 3-7 mm. Laser diodes with wavelength 0.67, 0.78; 0.81; 0.84, 0.85; 0.89; 0.905, 0.98; 1.06 μm are commercially available worldwide. These diodes correspond to the transparency window [42]. These diodes and collimator are used in apparatuses developed in our laboratory. In contrast to red and IR laser radiation, the depth of penetration of blue radiation is 0.5-1.0 mm. Combined (red + blue) laser radiation is effective in therapy of extremities, joints, skin diseases, etc. [11, 42]. The immune system can also be activated [11, 86]. Combined (red + blue, 1-min × 6 + 4 points in liver at radiation density 100-120 mW/cm^2) laser radiation activates bone synthesis at an injury site and cell proliferation. Synergistic effect is observed for laser radiation with power density 70-80 mW/cm^2 and magnetic field B = 50-75 mT [86]. Combined laser radiation exerts more pronounced effect on hepatocyte structure than red laser radiation [86]. It was demonstrated in this work that combined laser + magnetic therapy apparatuses should be developed. Such apparatuses should be based on solid-state lasers and nonlinear crystals. The size of the solid-state laser is a few millimeters. Variable wavelength apparatuses provide variation of radiation spot size in the patient's body. Lightguides provide radiation transport to internal pathological foci. Parameters of collimator 3, apparatus body 1, and tip 5 should be adapted to therapy purposes. Optimization of the tip parameters provides optimal therapy. Lightguides can also be used. The mobile collimator can be used for intravascular therapy.

3.3. Apparatus for Multicolor and Multifunctional Clinical Low-Intensity and Magnetolaser Therapy

Rodnik-1 (Table 1, Figure 9and Snag (Table 2, Figure 10) provide laser magnetotherapy using blue (0.47 μm), red (0.67 μm), near-IR (0.78; 0.81; 0.85; 0.89; 0.98; 1.06 μm) radiation sources and a mobile collimator All lasers except for the 0.89 μm laser work in continuous mode. Radiation power at the collimator output is 5 W, pulse time is ~100 nsec, pulse repetition frequency is 75-4800 Hz. Radiation power of continuous lasers is 1-500 mW. Laser radiation is modulated at 1-100 Hz; this option was described in [64, 87, 88].

Table 1. Technical characteristics of the Rodnik-1 magnetolaser therapeutic apparatus

Technical characteristic	Value
Number of independent channels	4
Wavelength, nm	450±30 670±20 780±20 890±20
Type of radiation source at wavelength^ 450 nm 670, 780, and 890 nm	 LED Laser diode
Operating regime with optical radiation at wavelength: 450, 670, and 780 nm 890 nm	 continuous pulsed
Adjustment range of the mean radiation power (with a step no greater than 1 mW) in the continuous regime, mW	10, for λ = 450 nm 0–25, for λ = 670 nm 0–30, for λ = 780 nm
Pulse width at the 0.5 level of the maximum, ns	100±50 (λ = 890 nm)
Pulse repetition rate, Hz	75, 150, 300, 600, 1200, 2400, 4800
Mean pulse power of the laser radiation, W	5
Maximum magnetic induction, mT	70
Operating regimes	manual, timer
Timer	from 1 sec to 60 min with a step of 1 sec
Mass, kg	no greater than 4

Continuous radiation was modulated at 50-70 Hz to increase the biological effect. At modulation frequency >100 Hz the biological effect of modulated radiation is similar to the biological effect of continuous radiation. Therapeutic effect does not depend on pulse repetition frequency (0.89 μm). Mean radiation power density of pulse lasers ($\tau = 100 \pm 50$ nsec) increases with pulse repetition frequency [42].

The physiotherapeutic apparatus developed in this work is based on a universal power source for four independent lasers. The block diagram of an apparatus for multicolor and multifunctional low-intensity laser and magnetolaser therapy is shown in Figure 11.

The consumer optionally determines the parameters of the lasers for the purposes of therapy. The optical power of the laser diodes is stabilized regardless of radiation temperature. Blood is exposed using red and IR lasers, as well as disposable sterile intravenous lightguides. Radiation power at the lightguide output is 1-5 mW. Blood intravenous laser exposure time is 15-20 min.

Table 2. Technical characteristics of the Snag magnetolaser therapeutic apparatus

Technical characteristic	Value
Wavelength of the radiation of the laser diode, nm: Snag 815 and Snag 812 Snag 855 and Snag 852 Snag 985 and Snag 982 Snag 1065 and Snag 1062	810±20 850±20 980±20 1060±20
Spectral range of radiation of the LED, nm	450±30
Regimes of action by optical radiation	continuous and modulated
Maximum power of laser radiation in the continuous regime, mW: Snag 815, Snag 855, Snag 985, Snag 1065 Snag 812, Snag 852, Snag 952, Snag 1062	500 200
Range of discrete regulation of the laser radiation power, % of maximum	10, 30, 50, 60, 70, 80, 90, 100
Modulation frequency, Hz	1, 2, 3, 5, 7, 10, 20, 50, 100, 200, 300, 600, 1000, 2000
Total radiation power of LED, mW	15
Magnetic induction, no less than, mT	20
Mass, no greater than, kg	1.3

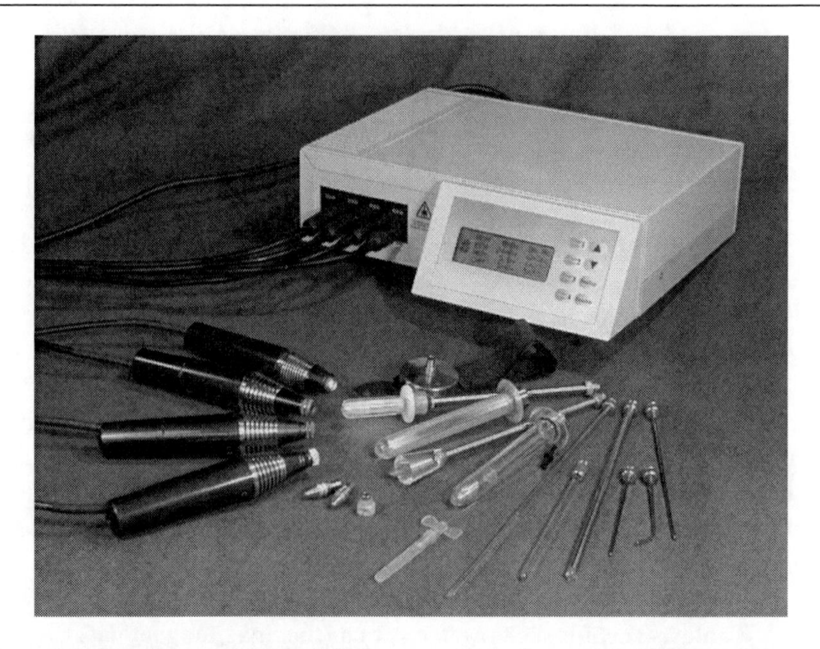

Figure 9. External view of the Rodnik-1 magnetolaser therapeutic apparatus.

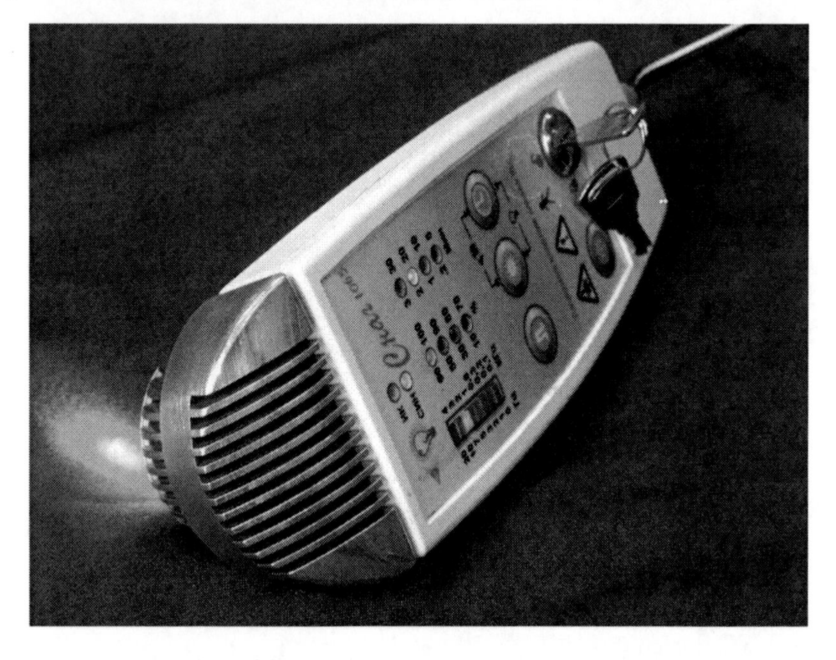

Figure 10. External view of the Snag magnetolaser therapeutic apparatus.

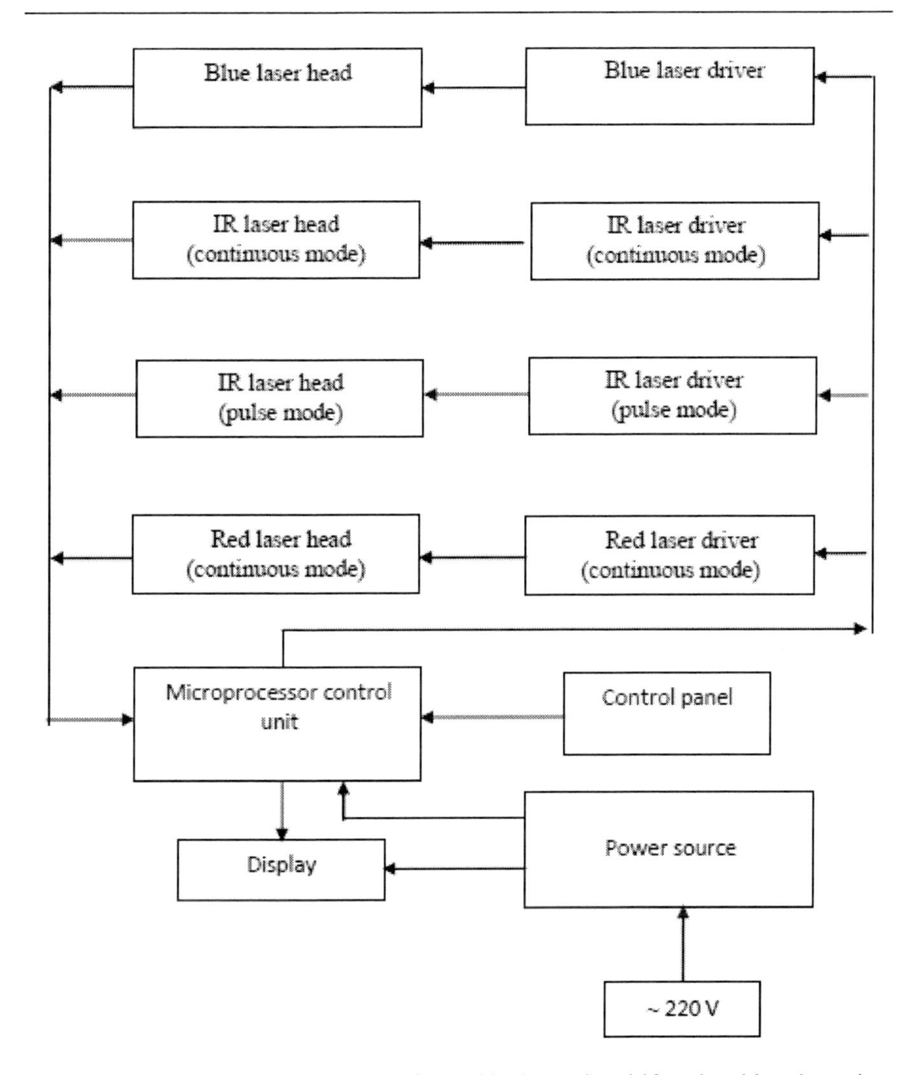

Figure 11. Block diagram of apparatus for multicolor and multifunctional low-intensity laser and magnetolaser therapy.

Therapeutic efficacy of the Rodnik-1 and Snag apparatuses lead to their use in clinics of the Republic of Belarus for blood exposure, photoblockage method, laser acupuncture, etc. [42]. The functional ability of the apparatuses is due to: a) variability of laser power within a broad range; b) lightguides of variable directivity; c) continuous, modulated, and pulse generation modes; d) variable radiation wavelength providing variable radiation penetration depth.

The use of the multifunctional lowintensity and magnetolaser therapy apparatuses in modern medicine [89] is advantageous in terms of social and economy factors because they reduce inpatient time and drug consumption without adverse side effects [42]. The economic effect was estimated in Belarus University and Municipal Hospital 6 of Minsk. This estimate demonstrated that red laser therapy reduced inpatient time from 173 ± 5 days to 156 ± 3 days. Combined blue + red laser therapy using the apparatus developed in this work reduced inpatient time from 173 ± 5 days to 141 ± 4 days.

CONCLUSION

The principles of construction of multi-functional (universal) multicolored therapeutic device were worked out. It provides the possibility of laser and magnetolaser exposure by all the methods agreed now in phototherapy, including: local and zonal magneto-laser therapy by external exposure; abdominal laser therapy, intra-articular and interstitial (the method of photoblockade) effect, intravascular (intravenous) effects on blood using disposable sterile fiber optic accessories with a spike; above-vessels (transcutaneous) magneto-laser effect on the blood, laser acupuncture (effects on biologically active points and zones) as well as combination of these technologies.

Multi-functionality of the device, implemented using modern element base of laser diodes, achieved through the developed by the authors technical solutions based on the use of a collimator, that was set with the capability of moving along the optical axis of the laser emitter.

Moving the collimator in relation to the laser diode that generates highly divergent radiation, allows either to adjust the size of the light spot on the patient's body by changing the divergence of the light beam or to focus the radiation on the input end of optical fiber monofilaments, used by intracavitary, interstitial and intravascular effects.

ACKNOWLEDGMENTS

The author is grateful to Professor V. S. Ulashchik, Professor N. S. Serdyuchenko, Professor P. S. Rusakevich, Dr. G. R. Mostovnikova, A.B. Ryabtsev, I. A. Leusenko, L. G. Plavskaya, A. I. Tret'yakova, A. V. Mikulich for fruitful discussions and critical remarks, and N.V. Bulko, I.V. Supeeva – for technical assistance in preparing this publication.

REFERENCES

[1] Inyushin, V M. (1965). *The Biological Activity of Red Radiation*. Alma-Ata (USSR): Kazakhstan; [in Russian].

[2] Inyushin, V M. (1967). *The Biological Effect of Monochromatic Red Light*. Alma-Ata (USSR): Kazakhstan; [in Russian].

[3] Inyushin, V M. (1970). *Laser Light and Living Organisms*. Alma-Ata (USSR): Kazakhstan; [in Russian].

[4] Inyushin, V. M., and Chekurov, P. R. (1975). *Biostimulation by Laser Beam and Bioplasma*. Alma-Ata, USSR: Kazakhstan; [in Russian].

[5] Mester, E. (1966). The use of the laser beam in therapy. *Orv. Hetil.* 107 (22), 1012–1016.

[6] Mester, E., Ludany, G., Sellyei, M. and Szende, B. (1968). On the biologic effect of laser rays. *Bull. Soc. Int. Chir.*, 27 (1), 68–73.

[7] Mester, E., Ludani, G. and Sellyei, M. (1968). The stimulating effect of low power laser rays on biological systems. *Laser Rev.*, 1 (1) 3–8.

[8] Mester, E., Ludany, G., Frenyo, V., Ihasz, M., Doklen, A., Jaszsagi-Nagy, E. and Tota, G.J. (1970). Experimental and clinical observations with laser rays. *Langenbecks. Arch. Chir.*, 327, 310–314.

[9] Mester, E., Spiry, T., Szende, B. and Tota, J.G. (1971). Effect of laser rays on wound healing. *Am. J. Surg.*, 122 (4), 532–535.

[10] Mester, E., Nagylucskay, S., Waidelich, W., Tisza, S., Greguss, P., Haina, D. and Mester, A. (1978). Effects of direct laser radiation on human lymphocytes. *Arch. Dermatol. Res.*, 263 (3), 241–245.

[11] Kryuk, A. S., Mostovnikov, V. A., Khokhlov, I. V. and Serdyuchenko, N. S. (1986). Therapeutic Effectiveness of Low-Intensity Laser Radiation. Minsk, USSR: Nauka I Tekhnika; [in Russian].

[12] Malevich, K. I., Gerasimovich, G. I. and Rusakevich, P. S. (1992). Methods of Laser Therapy in Obstetrics and Gynecology. Minsk, USSR: Vysheishaya Shkola; [in Russian].

[13] Illarionov, V. E. (1992). *Principles of Laser Therapy*. Moscow: Respekt; [in Russian].

[14] Pletnev, S. D. (1996). Lasers in Clinical Medicine. *A Handbook for Physicians*. Moscow: Meditsina; [in Russian].

[15] Moskvin, S. V. and Builin, V. A. (2000). *Low-Level Laser Therapy.* Moscow: NPLTs Tekhnika; [in Russian].
[16] Ohshiro, T. and Calderhead, R. G. (1988). *Low Level Laser Therapy: A Practical Introduction.* Chichester-New York-Brisbane-Toronto-Singapore: JohnWiley and Sons.
[17] Baxter, G. D. (1994). *Therapeutic Lasers: Theory and Practice.* Edinburgh: Churchill Livingstone.
[18] Tunér, J.and Hode, L. (2002). *Laser Therapy. Clinical Practice and Scientific Background.* Grängesberg: Prima Books.
[19] Tunér, J. and Hode L. (2004). *Laser Therapy Handbook.* Grängesberg: Prima Books.
[20] Durnov, L. A., Gusev, L. I., Balakirev, S. A., Grabovshchiner A. Ya., Ivanova Zh. V. (2000) Low-intensity lasers in pediatric oncology. *Vestn. Ross. Akad. Med. Nauk.* 6, 24-27; [in Russian].
[21] *Genot, M.-T. and Klastersky, J. (2005).* Low-level laser for prevention and therapy of oral mucositis induced by chemotherapy or radiotherapy. *Current Opinion Oncology*. 17, 236–240.
[22] Bensadoun, R. J., Franquin, J. C., Ciais, G., Darcourt, V., Schubertm M. M., Viot, M., Dejou, J., Tardieu, C., Benezery, K., Nguyen, T. D., Laudoyer, Y., Dassonville, O., Poissonnet, G., Vallicioni, J., Thyss, A., Hamdi, M., Chauvel, P. and Demard, F. (1999). Low-energy He/Ne laser in the prevention of radiation-induced mucositis: A multicenter phase III randomized study in patients with head and neck cancer. *Support Care Cancer* 7, 244–252.
[23] Parmon, E .M., Borshchevskii, V. V., Kamyshnikov V. S. and Bortkevich, L. G. (2003). Combined low-intensity laser radiation in renal tuberculosis. *Probl. Tuberk. Bolezn. Legk.* 6, 28–33; [in Russian].
[24] Kul'chavenia, E. V. (2001). Low-energy laser therapy in combined treatment of of patients with urinary tuberculosis. *Urologiia* 3, 8–11; [in Russian].
[25] Zinman, L. H., Ngo, M., Ng, E. T., New, K. T., Gogov, S. and Bril, V. (2004). Low-intensity laser therapy for painful symptoms of diabetic sensorimotor polyneuropathy: a controlled trial. *Diabetes Care*. 27, 921–924.
[26] Pinto, F. C., Chavantes, M. C., Pinto, N. C., Alho, E. J. L., Yoshimura, E. M., Matushita, H., Krebs, V. L. J. and Teixeira M. J. (2010). Novel treatment immediately after myelomeningocele repair applying low-level laser therapy in newborns: a pilot study. *Pediatr. Neurosurg*. 46 (4), 249–254.

[27] Bjordal, J. M., Couppe, C. and Ljunggren, A. E. (2010). Low level laser therapy for tendinopathy. Evidence of a dose-response pattern. *Phys. Ther. Reviews* 6, 91–99.

[28] Bjordal, J. M., Couppe, C., Chow, R. T., Tuner, J. and Ljunggren, E. A. (2003). A systematic review of low level laser therapy with location-specific doses for pain from chronic joint disorders. *Australian J. Physiotherapy*. 49, 107–116.

[29] Reddy, G.K. (2004). Photobiological basis and clinical role of low-intensity lasers in biology and medicine. *J. Clin. Laser Med. Surg*. 22, 141–150.

[30] Woodruff, L. D., Bounkeo, J. M., Brannon, W. M., Dawes, K. S., Barham, C. D., Waddell, D. L. and Enwemeka C. S. (2004). The efficacy of laser therapy in wound repair: a meta-analysis of the literature. *Photomed. Laser Surg*. 22, 241–247.

[31] Enwemeka, C. S., Parker, J. C., Dowdy, D. S., Harkness, E. E., Sanford, L. E. and Woodruff, L. D. (2004). The efficacy of low-power lasers in tissue repair and pain control: a meta-analysis study. *Photomed. Laser Surg*. 22, 323–329.

[32] Nakaji, S., Shiroto, C., Yodono, M., Umeda, T. and Liu, Q. (2005). Retrospective study of adjunctive diode laser therapy for pain attenuation in 662 patients: detailed analysis by questionnaire. *Photomed. Laser Surg*. 23, 60–65.

[33] Moshkovska, T. and Mayberry, J. (2005). It is time to test low level laser therapy in Great Britain. *Postgraduate Med. J.* 81, 436–441.

[34] Chow, R. T. and Barnsley, L. (2005). Systematic review of the literature of low-level laser therapy (LLLT) in the management of neck pain. *Lasers Surg. Med*. 37, 46–52.

[35] Posten, W., Wrone, D. A., Dover, J. S., Arndt, K. A., Silapun,t S. and Alam M. (2005). Low-level laser therapy for wound healing: mechanism and efficacy. *Dermatol. Surg*. 31, 334–340.

[36] Cowen, D., Tardieu, C., Schubert, M., Peterson, D., Resbeut, M., Faucher, C. and Franquin, J. C. (1997). Low energy helium-neon laser in the prevention of oral mucositis in patients undergoing bone marrow transplant: results of a double blind randomised trial. *Int. J. Radiat. Oncol. Biol. Phys.* 38, 697–703.

[37] Martin, R. (2003). Laser-accelerated inflammation/pain reduction and healing // *Practical Pain Management*. 3, 20–25.

[38] Brosseau, L., Wells, G., Marchand, S., Gaboury, I., Stokes, B., Morin, M., Casimiro, L., Yonge, K. and Tugwell, P. (2005). Randomized

controlled trial on low level laser therapy (LLLT) in the treatment of osteoarthritis (OA) of the hand. *Lasers Surg. Med.* 36, 210–219.

[39] Irvine, J., Chong, S. L., Amirjani, N. and Chan, K. M. (2004). Double-blind randomized controlled trial of low-level laser therapy in carpal tunnel syndrome. *Muscle Nerve*. 30, 182–187.

[40] Hopkins, J. T., McLoda, T. A., Seegmiller, J. G. and Baxter, G. D. (2004). Low-level laser therapy facilitates superficial wound healing in humans: a triple-blind, sham-controlled study. *J. Athl. Train.* 39, 223–229.

[41] Flemming, K. A., Cullum, N. A., Nelson, E. A. (1999). A systematic review of laser therapy for venous leg ulcers. *J. Wound Care*. 8, 111–114.

[42] Plavskii, V. Yu., Mostovnikov, V. A., Ryabtsev, A. B., Mostovnikova, G. R., Plavskaya, L. G., Nikeenko, N. K., Leusenko, I. A., Mostovnikov, A. V., Ginevich, V. V., Ulashchik, V. S., Rusakevich, P. S., Volotovskaya, A. V., Rybin, I. A. and Serdyuchenko, N. S. (2007). Apparatus for low-level laser therapy: modern status and development trends. *J. Opt. Technol.* 74(4), 246–257.

[43] Tunér, J. and Hode, L. (1998). It's all in the parameters: a critical analysis of some well-known negative studies on low-level laser therapy. *J. Clin. Laser Med. Surg.* 16, 245–248.

[44] Polonskii, A. K., Soklakov, A. I., Cherkasov, A. V., Nemtsev, I. Z. and Dreval' A.A. (1984). Experimental and clinical aspects of magnetolaser therapy. *Patol. Fiziol. Eksp. Ter.* 3, 49–52; [in Russian].

[45] Matyshova, MA, Aristova, VA, Golubenko, YuV and Dreval’ AA. Laser and magnetolaser therapy. In: Polonskii AK, ed. *Reviews of the major problems of medic*ine. Moscow: Medicine and Health Service; 1985; 1–65; [in Russian].

[46] Polonskii, AK. Laser and magnetolaser therapy. In: Pletnev SD, ed. *Lasers in Clinical Medicine*, Moscow: Meditsina; 1996; 378–39; [in Russian].

[47] Ulashchik, V. S. (2008). *Physiotherapy: Universal Medical Encyclopedia*. Minsk: Book House; [in Russian].

[48] Ulashchik, V. S. (2011). New methods of physical therapy and instruments for their application (based on projects developed in Belarus). *Vopr. Kurortol. Fizioter. Lech. Fiz. Kult.* 1, 28–32; [in Russian].

[49] Ulashcyk, V. S. and Volotovskaya, A. V. (2007). Current and long-term technologies of laser therapy. *Proc. SPIE.*, 6734.

[50] Fedorov, A. A., Riabko, E. V. and Gromov, A. S. (2010). The use of magnetic-laser therapy in the combined treatment of osteoarthrosis in workers exposed to inorganic fluoride compounds. *Vopr. Kurortol. Fizioter. Lech. Fiz. Kult.* 4, 20–22; [in Russian].

[51] Kneebone, W. J. (2009). Magneto-laser therapy of pulpitis and vertebra column osteochondrosis. *Practical Pain Management.* 9(8), 62–63; [in Russian].

[52] Gizinger, O. A. and Ishpakhtina, K. G. (2010). The study of expression of apoptosis receptors (CD95+) on the surface of neutrophils from cervical secretion of women with chlamydial infection and the possibility of its correction by magnetic laser radiation. *Vopr. Kurortol. Fizioter. Lech. Fiz. Kult.* 3, 29–31; [in Russian].

[53] Gaidashev, E. A., Lebedev, K. N., Khristoforov, V. N., Biriukov, V. V. and Gatkin, E. (1995). An evaluation of the effect of magnetic-laser therapy on external respiratory function in complicated forms of acute pneumonia in children. *Vopr. Kurortol. Fizioter. Lech. Fiz. Kult.* 3, 12–14; [in Russian].

[54] Korolev, Iu. N., Mikhaĭlik, L. V., Geniatulina, M. S. and Nikulina, L. A. (2010). The use of drinking sulfate mineral water in combination with laser and magnetic-laser irradiation for primary prophylaxis of post-radiation disorders (experimental study). *Vopr. Kurortol. Fizioter. Lech. Fiz. Kult.* 4, 3–6; [in Russian].

[55] Kruglova, L. S. (2008). Magnetic laser therapy in combined therapy of patients with atopic dermatitis. *Vopr. Kurortol. Fizioter. Lech. Fiz. Kult.* 1, 44–46; [in Russian].

[56] D'iakonov, A. V., Raĭgorodskiĭ, Iu. M. and D'iakonov, V.L. (2006). Magnetic-laser therapy with AMO-ATOS-LAST-LOR apparatus for a prevention of rhinosurgical complications of vegetative dysfunction. *Vestn. Otorinolaringol.* 4, 48–52; [in Russian].

[57] Prokhonchukov, A. A., Zhizhina, N. A., Banchenko, G. V., Milokhova, E. P., Saprykina, V. A., Kulazhenko, T. V., Aliab'ev, Iu. S., Semenova, L. L., Morozova, N. V., Vasmanova, E. V., Vakhtin, V. I., Mozgovaia, L. A. and Vinogradov, A. B. (2006). Prevention and treatment of face herpes with magnetic-laser radiation from Optodan unit. *Stomatologiia* (Mosk). 85(3), 78–82; [in Russian].

[58] Ponomarenko, G. N., Obrezan, A. G., Stupnitskii, A. A., Tishakov, A. Y., Kostin, N. A. (2005). Genetic determinants of efficiency of magnetic laser therapy of essential hypertension. *Bull. Exp. Biol. Med.* 139(3), 300–304; [in Russian].

[59] Iurshin, V. V. (2003). Optimization of magnetic laser therapy in the treatment of male infertility. *Voen. Med. Zh.* 324(8), 31–34; [in Russian].

[60] Mostovnikov, V. A., Mostovnikova, G. R., Plavskii, V. Yu., Plavskaya, L. G., Morozova, R. P. and Tret'yakov, S. A. (1991). On the mechanism of the therapeutic effect of low intensity laser radiation and a static magnetic field. In: Proc. Int. Conf. "New in Laser Medicine and Surgery", Moscow: Institute of Laser Medicine, 2, 192–194; [in Russian].

[61] Mostovnikov, V. A., Mostovnikova, G. R., Plavski, V. Y., Plavskaja, L. G. and Morozova, R. P. (1994). Primary photophysical processes which define the biological and therapeutic effect of low-intensity laser radiation. *Proc. SPIE*, 2370, 541–548.

[62] Mostovnikov V. A., Mostovnikova G. R., Plavskii V. Yu., Plavskaja L.G., Morosova R.P. (1995). Molecular mechanism of biological and therapeutically effect of low-intensity laser irradiation. *Proc. SPIE.* 2391, 561–573.

[63] Plavskii, V. Yu. and Barulin, N. V. (2008). Effect of polarization and coherence of low-intensity optical radiation on fish embryos. *J. Appl. Spectrosc*, 75(6), 843–856.

[64] Plavskii, VYu and Barulin, NV. Fish embryos as model for research of biological activity mechanisms of low intensity laser radiation. In: Arkin, WT, ed. *Advances in Laser and Optics Research*. New York: Nova Science Publishers. 2010; 4, 1–48.

[65] Plavskii V. Yu. and Barulin N. V. Investigation of biological activity mechanisms of low intensity optical radiation at the embryonic level. In: *Advances in Optics, Photonics Spectroscopy and Applications VI*. Hanoi. Vietnam: Publishing House for Science and Technology; 2011, 228–233.

[66] Plavskii V. Yu. and Barulin N. V. (2012). Effect of Polarization and Coherence of Optical Radiation on Sturgeon Sperm Motility. *World Acad. Scie. Engineer Technol.* 67, 947–951.

[67] Plavskii, V. Yu. (2010). Apparatus for Magneto-Laser Therapy with High Magnetic Induction in the Area of the Optical Radiation. *Biomed. Radioelektr.* (Moscow), 5, 63–68; [in Russian].

[68] Mostovnikov, V. A., Plavskii, V. Yu., Ryabtsev, A. B., Mostovnikova, G. R., Plavskaya, L.G., Mostovnikov, A. V., Leusenko, I. A. and Ginevich V. V. (2005). An Apparatus for Magnetophototherapy. Belarus Patent No. 2349; [in Russian].

[69] Plavskii, V. Yu., Mostovnikov, V. A., Ryabtsev, A. B., Mostovnikova, G. R., Plavskaya, L.G., Mostovnikov, A. V., Leusenko, I. A. and Ginevich V. V. (2005). *An Apparatus for Magnetolaser Therapy*. Belarus Patent No. 2392; [in Russian].

[70] Plavskii, V. Yu. (2011). Correction of Magnetic Field Distribution within the Optical Radiation Coverage Zone of Magnetic Laser Therapy Apparatuses. *Biomed. Engineering*, 45(1), 9–11.

[71] Plavskii, V. Yu., Ryabtsev, A. B., Leusenko, I. A., Mostovnikov, V. A., Mostovnikova, G. R., Plavskaya, L.G., Tret'yakova, A.I. and Mostovnikov, A. V. (2011). Principles of Development of Multifunctional Equipment for Low-Intensity Laser and Magnetolaser Therapy. *Biomed. Engineering*. 45(2), 54–58.

[72] Kertesz, I., Fenyö, M., Mester, E. and Bathori, G. (1982). Hypothetical physical model for laser biostimulation. *Opt. Laser Technol.*, 14 (1), 31–32.

[73] Fenyö, M. (1984). Theoretical and experimental basis of biostimulation by laser irradiation. *Opt. Laser Technol.*, 16 (4), 209–215.

[74] Kubasova, T., Fenyö, M., Somosy, Z., Gazso, L. and Kertesz, I. (1988). Investigations on biological effect of polarized light. *Photochem. Photobiol.* 48 (4), 505–509.

[75] Mostovnikov V. A., Mostovnikova G. R., Plavskii V. Yu., Plavskaya L. G., Morosova R. P., Tret'yakov S. A. (1992). The dependence of the biological activity of low-intensity laser radiation on the degree of polarization of the lightwave. Proc. Int. Conf. "Perspective directions of laser medicine". Moscow: Institute of Laser Medicine, 345–347; [in Russian].

[76] Kubasova, T., Horvath, M., Kocsis, K. and Fenyö, M. (1995). Effect of visible light on some cellular and immune parameters. *Immunol. Cell Biol.*, 73 (3), 239–244.

[77] Fenyö, M., Mandl, J. and Falus, A. (2002). Opposite effect of linearly polarized light on biosynthesis of interleukin-6 in a human B lymphoid cell line and peripheral human monocytes. *Cell Biol. Intern.*, 26 (3), 265–269.

[78] Plavskii, V.Y. and Barulin, N.V. (2008). The influence of infra-red laser radiation on viability of larva sturgeon fish to deficiency of oxygen. *Biomed. Radioelectronics* (Moscow), 8-9, 65–74; [in Russian].

[79] Plavskii, V.Yu. and Barulin, N.V. (2008). Effect of exposure of sturgeon roe to low-intensity laser radiation on the hardiness of juvenile sturgeon. *J. Appl. Spectrosc*, 75 (2), 241–250.

[80] Plavskii, V.Yu. and Barulin, N.V. (2009). Photophysical Processes that Determine the Biological Activity of Low Intensity Optical Radiation. *Biomed. Radioelektr.* (Moscow), 6, 23–40; [in Russian].

[81] Plavskii, V. Yu., Mostovnikov, V. A., Ryabtsev, A. B., Mostovnikova, L.G., Mostovnikov, A. V., Leusenko, I. A. and Ginevich V. V. (2007). An Apparatus for Laser Therapy. Belarus Patent No. 9405; [in Russian].

[82] Tanin, L. V., Nechipurenko, N. I., Vasilevskaya, L. A. Laser Hemotherapy in Treating Diseases of the Peripheral Nervous System. Minsk, *Magic Book*; 2004; [in Russian].

[83] Ryabtsev, G. I., Bogdanovich, M. V., Yenzhyieuski, A. I., Burov, L. I., Ryabtsev, A. G., Shchemelev, M. A., Pozhidaev, A. V., Matrosov, V. N., Mashko, V. V., Teplyashin, L. L. and Chumakou A. N. (2006). Parameters of the output beam of a longitudinally diode-pumped YVO_4/Nd:YVO_4-laser. *Quantum Electronics*, 36(10), 925–927.

[84] Batai, L. E., Kuz'min, A. N., Ryabtsev, G. I. and Demidovich A. A. (2000). Miniature diode-pumped Nd:YAG laser emitting in the blue. *J. Opt. Technol*, 67(11) 971–972.

[85] Batay, L. E., Demidovich, A. A., Kuzmin, A. N., Titov, A. N., Mond, M. and Kück, S. (2002). Efficient tunable laser operation of diode-pumped Yb, Tm:$KY(WO_4)_2$ around 1.9 μm. *Appl. Phys. B*. 75, 457–461.

[86] Serdyuchenko, N. S., Wrublevskii, V. A. and Archakova, L. I. (2009). The effect of laser radiation and the magnetic field on the maturation of the subchondral bone in artificially induced injury. *Photobiology and Photomedicine* (Kharkov), 6(4), 86–87.

[87] Plavskii, V.Yu. and Barulin, N.V. (2008). How the biological activity of low-intensity laser radiation depends on its modulation frequency. *J. Opt. Technol.*, 75 (9), 546–552.

[88] Plavskii, V.Y. and Barulin, M.V. (2009). Modulation effect of low-intensity laser radiation on its biological activity. *Laser Medicine* (Moscow). 13(1) 4–9; [in Russian].

[89] Ulashcyk, V. S., Volotovskaya, A. V. and Plavskii V.Yu. Combined magnetolaser therapy of patient with trophic ulcers and wounds, spinal osteochondrosis, vascular diseases of the joints. Instruction for use. Instruction No. 30-0406; Minsk: Ministry of Health of the Republic of Belarus; 2007; [in Russian].

In: Research Advances in Magnetic Materials
Editors: C. Toulson and D. Marwick
ISBN: 978-1-62417-913-6

Chapter 2

MAGNETIC NANOSTRUCTURES BY NANO-IMPRINT LITHOGRAPHY

Saibal Roy and Shunpu Li
Micropower-Nanomagnetics Group, Microsystems Center,
Tyndall National Institute, University College Cork,
Lee Maltings, Dyke Parade, Cork, Ireland

ABSTRACT

Patterned media, where the single domain recording bits are magnetically isolated from each other, offer the possibility to overcome many challenges faced by the other storage technologies, such as to obtain stable signal-to-noise ratio when the track width is reduced. The patterned media is likely to be the future media with high recording density and this has stimulated exploitation of different technologies to fabricate small featured magnetic materials. In parallel with the development of nanopatterning techniques there has been substantial progress in understanding and modelling the magnetic properties of small nanoparticles and patterned nano-dot arrays. The reduction of dimension leads to dramatic differences in micro/nanostructure, coercivity, anisotropy, and magnetic moment etc which attracted enormous attention from both academia and industry in many areas and spans far beyond the information storage field. Hence the paper is organised in the following way: In section I, as the patterned media is the most promising application for nanomagnet array, we briefly overview the data storage principle and storage media. Section II describes several alternative imprinting and pattern transfer technologies. Section III introduces the spin configurations, magnetization reversal properties of small magnets. Finally, in section IV a summary of current and possible future development in this area is outlined.

I. Data Storage Principle and Storage Media

Hard disk is a non-volatile storage device inside the computer that stores the data on it in the digital form. All the programs and data are stored on the hard disk. It is the most popular storage medium that is used for the desktop computer, laptop, video recorders and the gaming consoles. It is made up of the base, spindle motor, head set, voice coil and media. Data is recorded electromagnetically on the concentric circles or the tracks on the hard disk drive (HDD), which consists of several platters coated with the ferromagnetic material and each platter has read/write heads. Modern technology uses glass or ceramic platters. The magnetic layer on the platters has the magnetized domain, which is used to store the data that comes from the read/write heads. A schematic of a magnetic recording principle is illustrated in Figure 1.

To record and play back the information stored, one or more magnetic read/writeheads are used. The recording head consists of a high-permeability magnetic core with a narrow gap cut into it and a number of turns of a conductor wound around it. When electrical current flows through the conductor, magnetic flux emanate from magnetic core at the gap and penetrate the magnetic medium, causing the coated magnetic material to be magnetized to the right or the left (Figure 1a) direction. Binary data are encoded in the form of positive "1" or negative "0" in the magnetization in coincidence with a clock, which is synchronized with the disk or tape motion. With the advent of giant magnetoresistance (GMR), GMR stack readhead is nowadays being used to sense the magnetic flux emanating from the recorded transitions in the medium during read back. In order to achieve high recording density it is imperative that the head be very close to the medium. The storage capacity is usually measured by areal density, i.e. bits per inch. Areal density growth rate is frequently quoted as the measure of the speed of advancement of the technology. Since 1957 when IBM introduced the first magnetic disk drive with the RAMAC(Random Access Method of Accounting and Control), the areal density of magnetic disk recording has been increased over 2 million times by linear scaling of dimensions of the medium. While doing this, it has been necessary to reduce the size of the magnetic particles of which the medium is made in order to maintain the signal to noise ratio of the system because the signal to noise ratio scales approximately with the number of magnetic particles contained within a bit. The growing trend of the areal density in the past few years has slowed because of a fundamental limit in magnetic recording. This limitation relates to the fact that the magnetic material on the disk surface is necessarily composed of small grains. [1, 2] Because of the randomness of the grain shapes and sizes, each bit written on the disk need to cover about 100 grains to ensure that the information is reliably stored.

However, there is a lower limit to the size of a grain below which there is a risk that the magnetization may spontaneously reverse just due to thermal excitation universally present, even at room temperature. This phenomenon is called superparamagnetic transition. For a magnetic grain with volume V and anisotropy constant Ku, the anisotropy energy KuV will linearly reduce with its volume V, and eventually the energy is comparable with the thermal energy, i.e. $k_BT\sim KuV$ (k_B is Boltzmann's constant and T is the temperature). Under this situation the anisotropy no longer will be able to hold the magnetization in the direction of anisotropy. Because of such a competition of energies, stable magnetic bits for about 10 years, require a thermal stability factor KuV/k_BT larger than 60. [3] An alternative way to increase thermal stability of the nano-particles is to increase anisotropy of materials. But the high Ku implies that higher writing field is needed because the switching field, Hc, is proportional to Ku and inversely proportional to the saturation magnetization Ms of the medium ($Hc\sim 2Ku/Ms$ for single domain nanoparticle according to Stoner-Wohlfarthmodel). Being intrinsic property for a particular composition there is very limited scope to increase Ms, therefore it is believed that the longitudinal recording, which has been dominant technology for the past 50 years, is expected to be phased out soon.

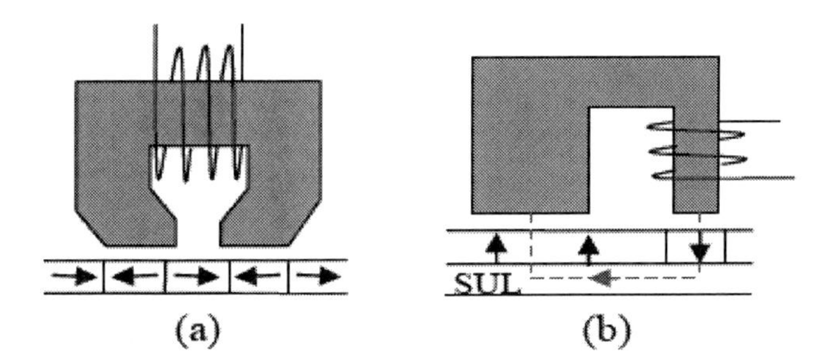

Figure 1 Schematic illustration of magnetic recording principle: longitudinal (a) and perpendicular recording (b).

Perpendicular magnetic recording (PMR) on the other hand addresses this thermal limit issue and allows continued advances in areal density. [4, 5] Unlike longitudinal recording where the magnetization in the bits is aligned circumferentially along the track direction, in the case of PMR the bits point up or down in the normal direction of the disk. As illustrated in Figure 1b a single pole PMR head combined with a soft-under-layer (SUL) offers a strong perpendicular writing field. For the PMR writing, the magnetic field is generated from the pole surface and collected by the SUL unlike the LMR writing, where the field is generated from the gap. Therefore, for PMR technology materials with higher *Ku* can be used. Because of this advantage of PMR technology it has been extensively researched in the past years and significant progress was made, and PMR technology drive has also been realised. [6] Thetypical magnetic material used in such media is CoCrPt alloys with combination of oxide like SiO_2. The estimated areal density for such PMR media is about 600Gb/in^2. To further increase areal density of PMR technology by using perpendicular media having much increased anisotropy, such as FePt film, [7, 8] one can increase the temperature of the media to reduce the field *Hc* needed for magnetization reversal. Based on this principle a heat-assisted magnetic recording (HAMR), a technology that magnetically records data on high-stability media using laser thermal assistance to first heat the material, has been developed. [9, 10] HAMR was developed by Fujitsu in 2006 so that potentially it could achieve one terabit per square inch densities. However since it faces many challenges, such as head integrationetc, [11] fully functional HAMR head is yet to come in the market.

II. Patterned Media and Nanoimprint

A promising method that has been considered for about decade is patterned recording media. [12-14] This technology abandons the idea of multiple grains for each bit. Instead single domain magnetic islands are used, and therefore allow for larger volume magnetic entity than the individual random grain used in continuous magnetic media. There are many challenges for creating data recording system based on patterned media, such as development of cost-effective high volume patterning techniques and technology bottleneck for accurate individual addressability of each patterned bit. Table 1 shows typical bit size for different required areal density. For instance, the size and periodicity of the islands required for 1Tb/in^2 bit density is 18nm and 25nm.

Table 1.

Areal density (Gb/in^2)	Island size (nm)	Periodicity (nm)
300	30	45
500	25	35
1000	18	25
1500	14	20

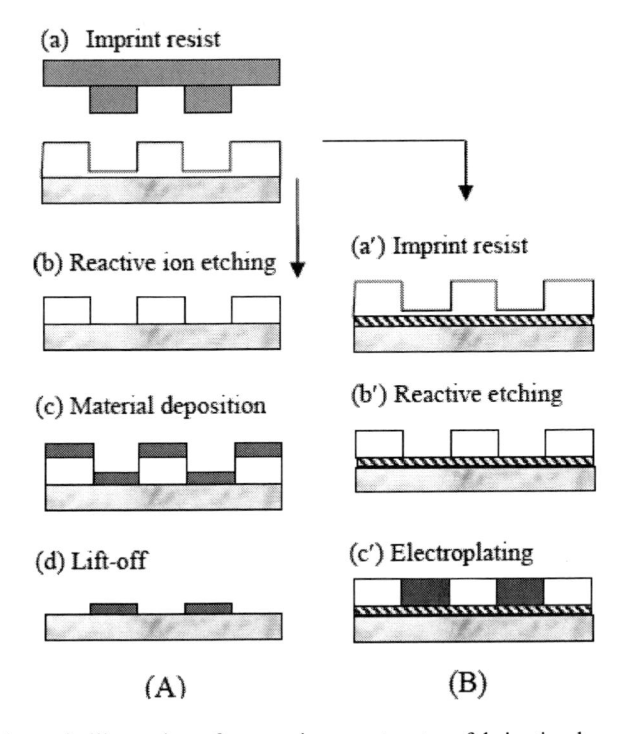

Figure 2. Schematic illustration of magnetic nanostructure fabrication by nanoimprint with different pattern transfer process (A) lift-off and (B) electroplating.

Currently, the only conventional lithographic technology capable of high throughput for fabrication of tens of millions of disks is optical lithography. However, the pattern densities shown in the table are generally beyond the reach of optical patterning.

One emerging technology showing significant promise in achieving both high throughput and resolution at reasonable cost is nanoimprint lithography. [15-18] In this technology topographic patterns are replicated from master mould into a polymeric resist coating on a disk substrate using a moulding-like process. A general nanoimprint and pattern definition process is shown in Figure 2. There are two ways to perform the nanoimprint: thermal or ultraviolet (UV) imprint depending on how the resist that used for imprint is cured. For the thermal imprint a layer of polymer resist is spin-coated on a substrate, then, a prefabricated stamp is brought to imprint the resist under a temperature well above the glass transition point of the resist where the resist will flow. After cooling the sample down to room temperature the stamp is released and a complementary pattern is defined in the resist layer (Figure 2A(a)). After application of reactive ion etching (RIE) to entirely etch the resist layer through (Figure 2A(b)) a magnetic film is deposited by sputtering or thermal evaporation etc process. (Figure 2A(c)).

Finally, a lift-off process is applied by removing the material on the top of resist through dissolving the resist in a solvent to define a final magnetic pattern (Figure 2A(d)). Alternatively, other deposition technique, like electroplating, can also be applied as shown in Figure 2B. In the case of electroplating a conductive seed layer need to be deposited before resist deposition. The UV-imprint is similar to the thermal imprint, except that the coated resist film is liquid which is cured by UV light after imprinting the stamp into the resist. [18] In this case either the stamp or the substrate need to be UV transparent. The UV imprint has advantage over the thermal imprint because no heating is required and hence the process is faster.

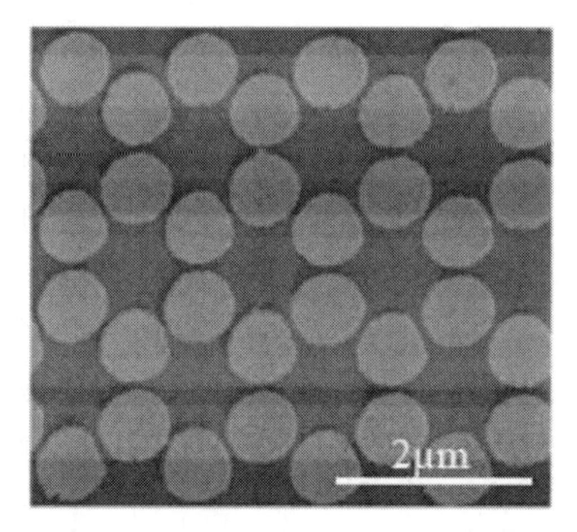

Figure 3. A permalloy structure defined by nanoimprint . Reprinted with permission from M. Natali et al. J. Appl. Phys. 96, 4334, 2004 [19]. Copyright 2012 by American Institute of Physics.

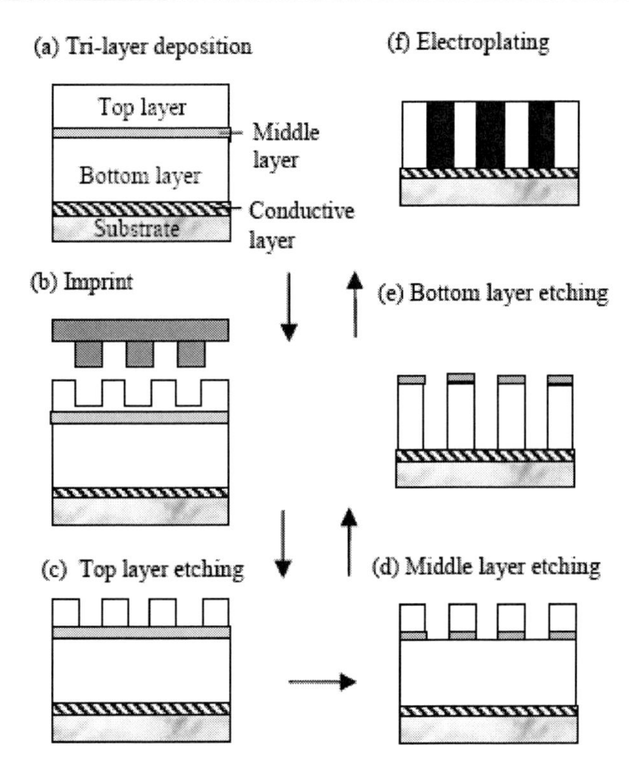

Figure 4. A illustration of trilayer process for fabricating nanostructure with high aspect ratio.

The stamp for imprinting can be any wear-resistant materials such as oxide, metal etc. Anti-adhesion treatment is needed for separating the stamp easily from samples without damaging the patterned structure. Figure 3 shows a typical permalloy structure defined by nanoimprint.

One problem of the above described process is aspect ratio which directly affects the quality of the patterned magnetic structure. For example if the lateral dimension of magnetic structure is ~100nm, the resist thickness will be in the similar dimension ~100nm(for easy and complete pattern transfer). Thus the thickness of magnetic material needs to be less than 30nm(about one third of resist thickness for lift off). The aspect ratio becomes even more important when a vertical pillar structure is electroplated. A trilayer approach is favourable in obtaining a pattern with higher aspect ratio. [20] Tri-layer structure consists of bottom/middle/top layers that is sequentially deposited on a substrate (Figure 4a). The top layer is imprinting layer on which a nanostructure is generated by imprinting. The bottom layer is a stable polymer layer which could not be damaged by the process of top layer. The middle layer acts as a pattern transfer layer which can be dry etched selectively using standard semiconductor process, for instance a few nanometer thick germanium etc.

After creating a nanostructure in top layer by imprint (Figure 4b) a dry etching is applied to etch through the top polymer layer entirely (Figure 4c). Then adifferent plasma is used to etch through the middle layer (Figure 4d), and subsequently switch the plasma back to etch through the bottom polymer layer by using the middle layer as mask (Figure 4e). Finally, the nanomagnets with high aspect ratio are deposited by electroplating (or other method) (Figure 4f).

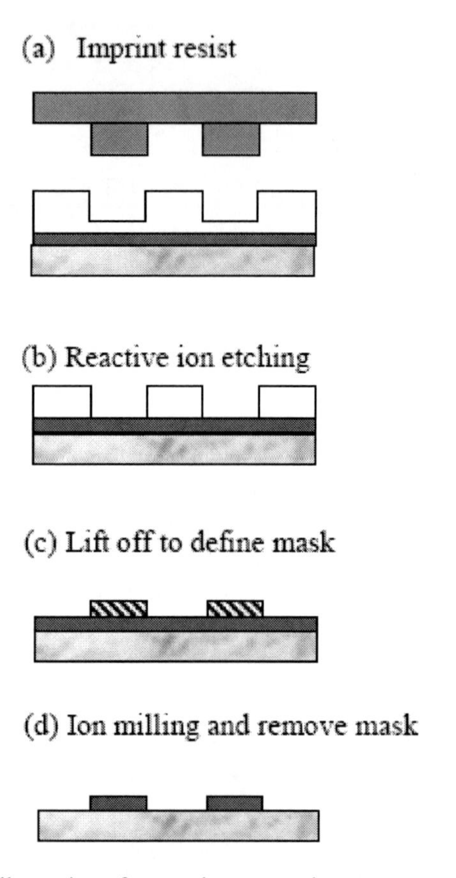

Figure 5. Schematic illustration of patterning magnetic nanostructure by imprint and ion-milling.

(a) Dispense imprint resist by scanning inkjet head

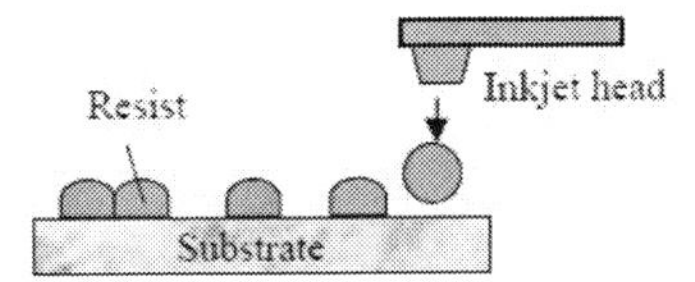

(b) Position the imprint template/stamp

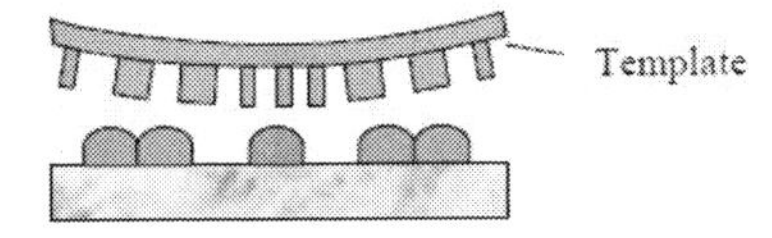

(c) Imprint and UV curing

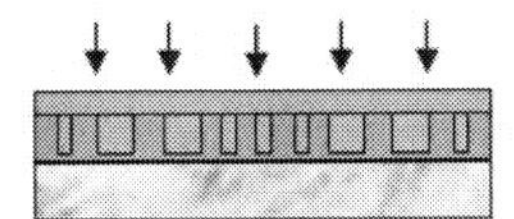

(d) Release template

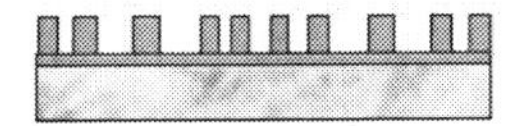

Figure 6. Schematic illustration of J-FIL process.

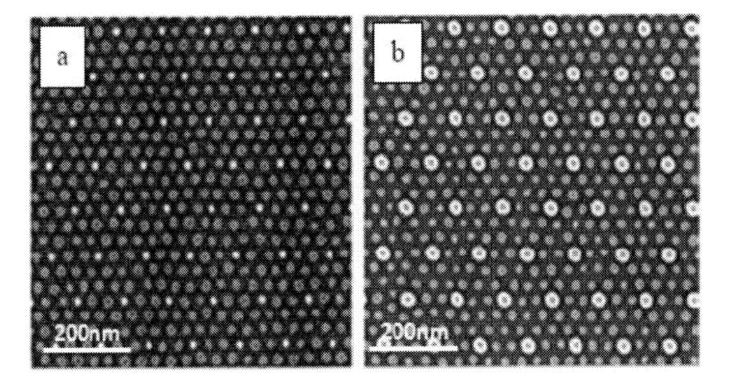

Figure 7. SEM images of ordered BCP[polystyrene-b-polydimethylsiloxane (PS-b-PDMS)]structure formed within a sparse 2D lattice of HSQ posts (brighter dots). The ordered PDMS phase is embedded in PS matrix. The substrate (Si) and post surfaces were functionalized with a PDMS brush layer in (a) and with a PS brush layer in (b). Reprinted with permission from Ion Bita et al. Science 321, 939, (2008)[25]. Copyright 2012 by Science.

Another important way of transferring pattern generated by nanoimprint is ion-milling which is particularly for patterning predeposited films, such as multilayered magnetic film (Figure 5). A polymer layer is spin coated on the predeposited magnetic layer. Then an imprint, dry etching, and lift-off process (Figure 5a-c) is done to define a pattern of mask for ion milling (as described in Figure 2A). The magnetic bits are then formed by ion-milling. The frequently used mask for ion-milling is aluminium film thanks to the readily formed thin layer of aluminium oxide which is highly resistive to ion-milling.

Among many alternative imprinting technologies developed so far, the Jet and flash imprint lithography (J-FIL), a form of UV-imprinting, has been demonstrated asa very suitable type of nanoimprint technique for patterned media and semiconductor applications (Figure 6). [21] The J-FIL is performed at room temperature and low pressure UV resist curing process. In J-FIL, a low viscosity UV-curable liquid resist is dispensed using an ink-jet process. A transparent template/stamp is then pressed into the resist such that the resist fills the pattern in the template. A UV light is used to cure the resist and the template is released. J-FIL provides additional benefits as follows: (i) J-FIL system includes a self-contained material dispense module which eliminates the separate material-dispensing spin coating system required by spin-on UV or thermal imprint processes, saving millions in capital and valuable clean room floor space. (ii) The amount of materials is controlled locally according to the structural density, thus overcome the problems of materials shortage/over-accumulation faced when spin-coating is used.

After more than a decade of development, the nanoimprint technology has now become a powerful technique to generate nanostructure over large areas at low cost and high throughput. A large area pattern with feature size down to sub-15nm using UV nanoimprint lithography has been demonstrated recently, [22] which is promising for low cost HDD technology.

Further a combination of nanoimprint and self-assembling of block copolymer (BCP) might provide a new solution to achieve high areal density of hard disk.

Use of self-assembled block copolymer as template to fabricate high density magnetic nanostructure has been demonstrated by Thurn-Albrecht et al [23] and recently T Ghosal et al. [24] In this technique, diblock copolymer is used to form a self-assembled template of dense periodic arrays with lattice constant typically 5~50nm. This opens a new avenue to fabricate hard disc with density higher than 1 Terabit per square inch (Tb/in^2).

However, the spontaneous process leads to the formation of “polycrystalline” micro-domain arrays consisting of randomly oriented regions, or grains, which limits the potential application. A workable patterned media recording system requires a highly uniform magnetic island array.

Recently, new method has been found to eliminate defects and impose long-range order of BCP by using a lithography generated guide pattern on substrate with lower resolution. [25-27]

The lattice constant of guide pattern (L_g) can be several times larger than lattice constant of BCP (L_b). Figure 7 shows such highly ordered nanodot array of BCP [polystyrene-b-polydimethylsiloxane (PS-b-PDMS)] guided by 2D array of hydrogen silsesquioxane(HSQ) posts fabricated by electron beam lithography.

Such resolution duplication technique opens a new way to realize high density patterned media by combination of nanoimprint technique with self-assembling of block copolymer. For 1Tb/in^2 patterned media the lattice constant of pattern is ~25nm, i.e. if we choose L_g=3L_b, then L_g~75nm is required for guiding pattern which is in the technically matured patterning scale of nanoimprint lithography.

III. Magnetic Properties of Patterned Magnetic Nano Structure

There has been substantial progress in understanding magnetic properties of magnets with reduction of size. [28-33] It is unfavourable if domain walls exist within a magnetic storage element. If the element is polycrystalline the grains should be exchange coupled and a single domain magnetization configuration is highly desired. Fortunately, this is the case when the size of magnetic element goes down to nanometer scale as a consequence of energy minimization. Although both perpendicular and in-plane magnetized nanomagnets can be used for storage, the magnetism of in-plane magnetized nano-structures is more complicated and informative from nanomagnetism point of view. In this section we review the magnetization configuration and reversal property of small magnets with in-plane magnetization.

(1) Size-Dependence of Magnetic Structures

The magnetic structure is controlled by the total energy available, which generally includes exchange energy (E_{ex}), crystalline anisotropy energy (E_a), magnetostatic energy (E_{ms}), and Zeeman energy (E_{zm}).

$$E_{tot}= E_{ex}+ E_{ms}+ E_a+ E_{zm}, \quad (1)$$

The magnet will adopt a particular type of magnetic structure depending on the result of the total energy minimization.

For a small disk shaped magnet with a weak crystalline anisotropy and without external field the contribution from last two terms in the equation (1) is small, and the magnetic structure is mainly defined by exchange energy and magnetostatic energy. The exchange energy and magnetostatic energy are expressed by relation: $E_{ex}=-\Sigma A_{ij}\boldsymbol{S}_i.\boldsymbol{S}_j$ and $E_{ms}=\mu_0 NM^2/2$ where the A and N are exchange constant and demagnetizing factor respectively. The minimization of exchange energy aligns the spins parallel to each other in a ferromagnet, i.e. single domain state.

However the single domain state will create magnetic charge on the edge of the small magnets and in turn increase the magnetostatic energy. For a micrometer scaled magnet, to reduce the total energy the magnet often split into multidomains [34] (Figure 8a), wherethe multidomain structure minimizes magnetostatic energy with a moderate increase in exchange anisotropy.

Two types of multi-domain structure can be found depending on the materials. For instance, if the crystalline anisotropy is strong, such as in epitaxial films clear domain walls can be viewed. [28, 33] While in material with weak crystalline anisotropy multi-vortex structure often exists.

With further reduction of size (100nm~1μm) a single vortex configuration is more energetically favoured than multidomains [30, 32] (Figure 8b). When the magnet size is reduced further, the exchange energy of vortex is increased drastically and a single domain state becomes a stable state [30, 32] (Figure 8c).

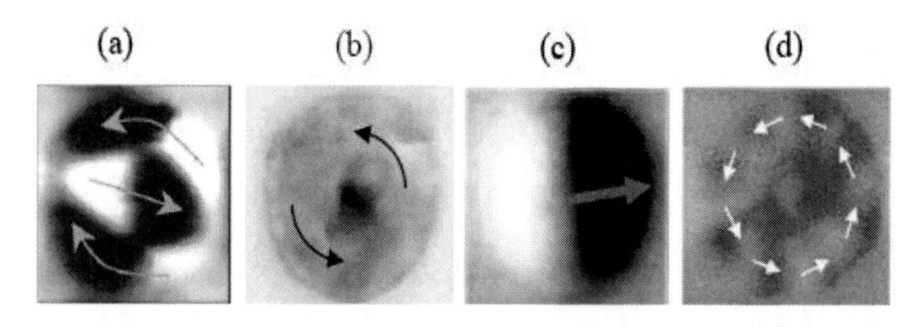

Figure 8. MFM images of magnetization configuration in small magnets. Reprinted with permission from M. Natali et al. Phys. Rev. Lett., 88, 157203, (2002) and S. P. Li et al. Phys. Rev. Lett. 86, 1102, (2001) [32, 34]. Copyright 2012 by American Physical Society.

For applications, such as storage and magnetic logic etc, [35, 36] an anisotropy is required for the single domain magnets to hold the giant spin in a global easy direction, and this can be done by generating a crystalline or induced easy axis during the material deposition or by patterning the structure into short bar or pillar shape to generate a shape anisotropy. [37] Ring shaped magnets has also been brought into attention in the past decade (Figure 8d). The ring shaped magnet is more stable than vortex as the exchange energy is largely concentrated in the center of vortex in a disk magnet which is removed by adopting a ring shape. [33] The two magnetization configurations (i.e. clockwise and anticlockwise) in a ring magnet can be used as two stable digital states for storage application.

Many factors can influence the detailed magnetic structure in a small magnet. For instance a thinner disc shaped element is favourable for a single domain state as thick element will intend to adopt vortex state due to the increased magnetostatic energy. [30] A stronger intrinsic anisotropy (crystalline anisotropy or induced anisotropy by film deposition etc) can stabilise single domain configuration, because the vortex state will increase the energy of anisotropy. [34] The element shape can also influence the spin configuration in the patterned element, in particular for soft magnetic material. For instance "S" or "C" shapes of spin-configurations can be found in a near-single-domain state. [38]

The vortex configuration is the consequence of shape confinement when the size of magnet goes to mesoscale: all spins intend to align along the edge of element to minimize the magnetostatic energy and the domain wall is "compressed" into the centre of the element. For in-plane magnetized small magnets the exchange energy is drastically increased in the center of the magnet due to the increase in angle between neighbouring spins. To decrease the exchange energy the magnetization becomes perpendicular to the film plane at the centre of the vortex. [39] Depending on the size of the element the contour of vortex can be strongly modified and multiple vortices with same or different chiralities can also be found in a single small element. [38, 40]

(2) Magnetization Reversal

The Stoner-Wohlfarth(SW) model is the simplest model that describes adequately the physics of single nanomagnets.[41] A nanomagnet with a uniaxial anisotropy is subjected to an external applied static magnetic field H (Figure 9). θ is the angle that M makes with the easy axis, while α is the angle that the applied external magnetic field makes with the easy axis. The total energy consisting of various anisotropy energies (like crystalline anisotropy and shape anisotropy), and Zeeman energy can be expressed as:

$$E = [K_C + K_S]sin^2\theta - HM_S\cos(\alpha - \theta) = K_E sin^2\theta - HMs\ cos(\alpha - \theta) \quad (2)$$

where K_C and K_S are crystalline and shape anisotropy constant, respectively. While $K_E=K_C+K_S$ is an effective anisotropy constant which includes contributions from crystalline anisotropy, shape anisotropy, and other anisotropies. At equilibrium condition the magnetization points along a direction defined by an angle θ^* that minimizes the energy. The minimum energy condition at $\theta^*$is:

$$\left(\frac{\partial E}{\partial \theta}\right)_{\theta=\theta^*} = 0 \text{ and } \left(\frac{\partial^2 E}{\partial \theta^2}\right)_{\theta=\theta^*} \geq 0, \text{ and we have}$$

$$[2K_E sin\theta cos\theta - HM_S\sin(\alpha - \theta)]_{\theta=\theta^*} = 0 \quad (3)$$

$$[2K_E(\cos^2\theta - sin^2\theta) + HM_S\cos(\alpha - \theta)]_{\theta=\theta^*} \geq 0 \quad (4)$$

For general α, the above equations cannot be solved analytically, except for $\alpha=0$, $\pi/4$, and $\pi/2$. Here we provide the case study for the most frequently encountered hysteresis loops: longitudinal easy and hard axis loops (α=0 and π/2).

(i) In the case α=0, equation (3) and (4) has solution $\theta^*=cos^{-1}(-HM_S/2K_E)$ and $HM_S/2K_E= \pm 1$ when $H \leq 2K_{eff}/M_S$ otherwise $\theta^*= 0,\pi$ yielding a square hysteresis loop. The measured magnetization $M_{//}= M_S cos(\alpha-\theta)= \pm M_S$

(ii) In the case $\alpha=\pi/2$, equations (3) and (4) give $\theta^*=\sin^{-1}(-HM_S/2K_E)$ when $H \leq 2K_{eff}/M_S$ otherwise $\theta^*=\pi/2$ yielding the measured magnetization $M_{\parallel} = \frac{M_S^2}{2K_E}H$, which is the main diagonal hysteresis loop.

Figure 9 shows a schematic illustration of easy and hard axis hysteresis loops for a single domain nanoparticle.

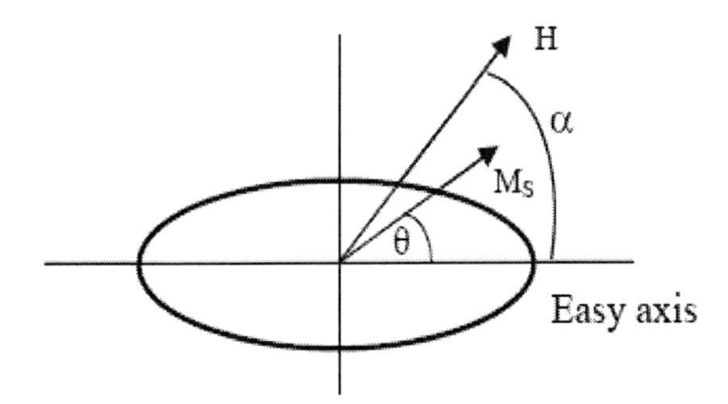

Figure 9. Coordinate system for magnetization reversal process in single-domain particle in which the shape and crystallographic easy axes coincide.

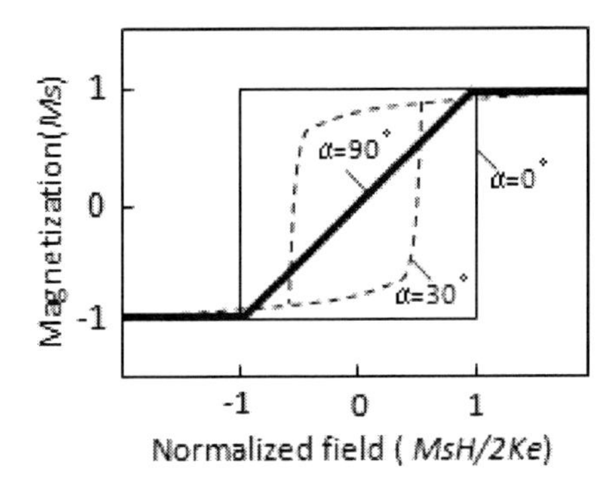

Figure 10. A schematic illustration of hysteresis loop of a nanoparticle for field direction in angle α=0°, 90° and 30°.

The square hysteresis loop in easy axis indicates that the reversal of magnetization is controlled by coherent rotation. The coercivity of easy axis loop equals the anisotropy field (saturation field of hard axis loop), i.e. $H_C = H_K = 2K_E/M_S$. The more details description of the SW model can be found from references [40].

The magnetization reversal of single domain and vortex nanomagnets has been extensively studied by using various techniques, such as magneto-optical Kerr effect, [30, 32, 33] micro-SQUID, [43] Kerr microscope, [44] magnetic force microscope (MFM). [28, 45] Figure 11 and Figure 12 show reversal processes of the magnetization of 500nm diameter Co dots with 30nm and 10nm thickness, respectively. For the 30nm thick dot as shown in Figure 11 under large negative values of the external field, the dots with an in-plane single domain configuration (i) with a characteristic black and white contrast.

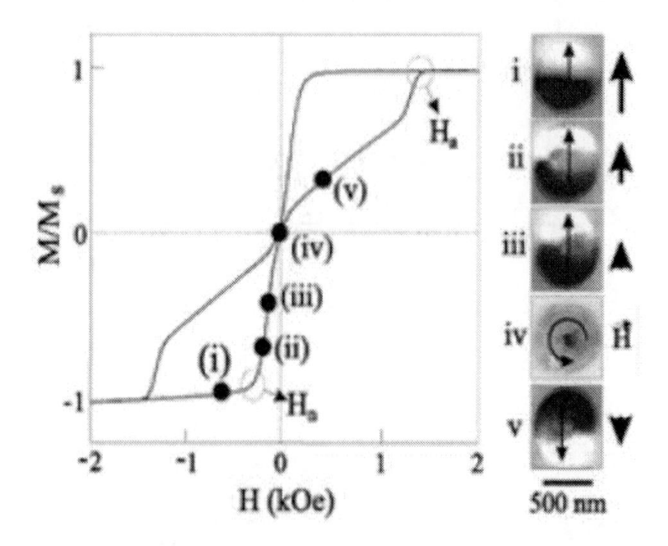

Figure 11. Reversal of the magnetization through vortex formation for a 30nm thick and 500nm diameter dot. Different points on the hysteresis curve (from i to iv) are visualized by MFM. Reprinted with permission from I. L. Prejbeanu et al. J. Appl. Phys. 91, 7343, (2002). [45]. Copyright 2012 by American Institute of Physics.

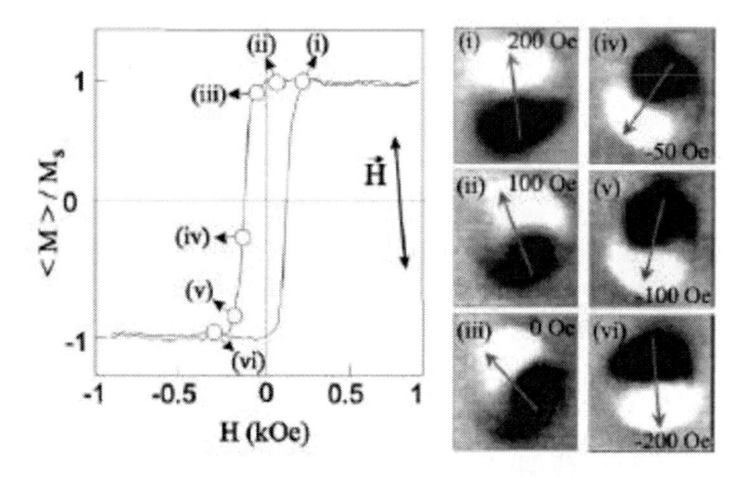

Figure 12. Reversal of the magnetization by coherent rotation for a 10nm thick and 500nm diameter dot. The MFM images (from i to vi) display the magnetic contrast corresponding to the distribution of the magnetization at different applied filed values. Reprinted with permission from I. L. Prejbeanu et al. J. Appl. Phys. 91, 7343, (2002). [45]. Copyright 2012 by American Institute of Physics.

With increasing the field, the net magnetization drops at the nucleation field Hn, where a vortex nucleates at the dot border (ii). By further increasing the field, the vortex moves towards the dot center (iii) and a completely symmetric configuration is stabilized at remanent state(iv), where the magnetization does not produce any magnetic stray field, and the only contrast arises from the vortex core is at the centre of the dot. Upon increasing the field, the net magnetization progressively reappears as the vortex core moves towards the opposite dot border, giving rise to a dipolar contrast that gradually develops, as shown in (v). Upon increasing the field above the annihilation field Ha, the vortex is expelled from the dot and the single domain state is again established. For the 10nm thick dots the single domain is the stable remanent state and reversal process is shown in Figure 12. The first image, recorded at 200 Oe, shows a uniformly magnetized dot in the field direction. The magnetization then tilts away from the field direction with reduction of the magnetic field. At -50Oe field (iv) the MFM image contrast is reversed, and finally a negative saturation is obtained at -200Oe field.

(3) Magnetostatic Interaction

Magnetostatic interactions in coupled nanomagnets have attracted much attention in recent years as they affect the areal density in hard disk application and also potentially useful for future magnetic random access memories and logic devices. [38] It is observed that magnetic particles in an array exhibit a range of switching field. [46, 47] Apart from an intrinsic variability between particles owing to small difference in their shape, size, or microstructure, a significant contribution to the spread in switching field is magnetostatic interactions between particles.

If the anisotropy of nanoparticles is not sufficiently high the maximum interaction field on a particle from its neighbours may become greater than its coercivity. Thus the particles can spontaneously reverse through interaction from their neighbours, and such patterned array would be unsuitable for data storage.

The switching field distribution exist for both perpendicular and in-plane magnetized pattern array, however the interaction induced reversal mechanism are different in these two cases. For the perpendicularly magnetized well-isolated dot array the measured hysteresis loop appears to be square, while for the close spaced array the hysteresis loops becomes sheared with a slop that is proportional to the interaction strength. In contrast, the inter-dot interaction in an array with in-plane magnetization does not induce a sheared loop and the loop is square. The difference of the two types of hysteresis loops can be schematically explained in Figure 13. At the initialstage of magnetization reversal in an array, the reversal field of a dot (*Hs*) is defined by a difference between intrinsic switching field (H_{in}— a field required for switching an isolated dot) and interaction field (*h*) from neighbours(H_{in}-h), because at this stage the interaction field assists the dot to reverse (lower panel in Figure a). In contrast, the reversal field at finishing stage of the reversal H_f is defined by $H_{in}+h$ due to the stabilization effect of surrounding dots (upper panel in Figure 13a). For the in-plane magnetization (IPM), the situation is more complicated than a dot array with perpendicular magnetization (PM). If all particles are magnetized parallel to each other, then the interaction field from some neighbours are demagnetizing, while the field from others stabilize the magnetization state of dots.

One can simply understand the reversal behaviour of the IPM dot array by examining nearest neighbours of a dot in the applied field direction because of their dominant contribution (Figure 13b). At the initial stage, the interaction field from neighbours stabilizes the magnetization (lower panel of Figure 13b) and the reversal field H_s is higher than the field $H_{f,}$ the field needed for switching a dot at finishing stage where the interaction field from nearest neighbours destabilize the dot.

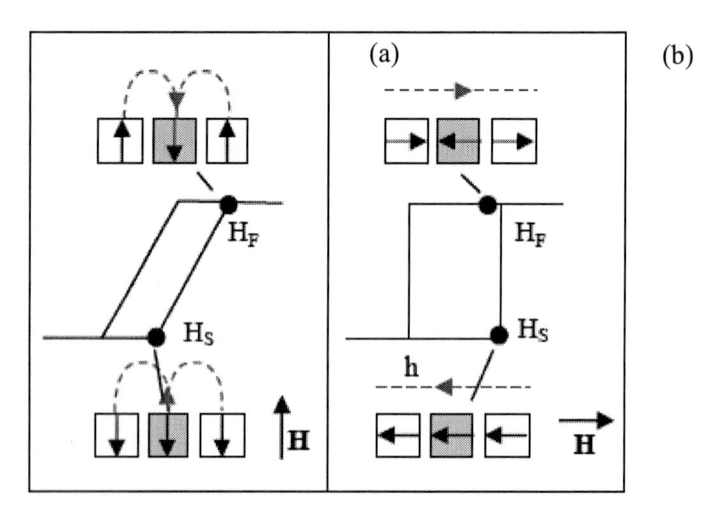

Figure 13. Schematic illustration of magnetization reversal of patterned dot array with perpendicular and in-plane magnetization.

IV. FUTURE DEVELOPMENT

As stated above that the challenge of increasing areal density of hard disk drive is to overcome the superparamagnetic transition problem. To satisfy $KuV >> k_B T$, one can see two approaches for extending the superparamagnetic limit of storage density to 1Tb/in^2 and beyond.

One approach is to increase Ku and another way is increasing V, the volume of individual magnetic particle. For the first approach a further downscaling of grain size of magnetic material of continuous media with engineered high anisotropy is required. The high anisotropy media is limited by the lack of disruptive technology that can dramatically increase the obtainable magnetic field in writing head presently. Alternative solution to increasing field of the writing head is to integrate a heating device, which is able to heat the media locally to reduce its switching field, with the head. The challenge faced in developing such thermally assisted magnetic recording is the design and fabrication of the writing head and developing suitable media. [48-51] The required media should havehigh anisotropy, small grain size (~4nm) with very narrow distribution, high thermal conductivity to cool down the written bits while maintaining a sharp thermal gradient for elevating local temperature for switching the bits.

For the second approach the media is patterned into single domain isolated magnetic bits. Current available nanofabrication technologies which offer well defined patterned array with areal density higher than 1 Tb/in^2 are e-beam and nanoimprint lithography. As the e-beam lithography offers very low throughput which cannot meet the required demand of mass production of hundreds of disks per hour, the nanoimprint appears to be the most promising technique.

The single layer patterning, circular symmetry of the disks eliminate the need of aggressive accuracy requirement of alignment. Fortunately, the nanoimprint lithography has received a lot of attention particularly in optical and electronic industry lately, where the industry is targeting 22nm node flash memory production in 2013 and aiming to further reduce to 11nm in 2019.and many of those concepts could be used in HDD industry.

There are two types of patterned media under development—bit patterned media (BPM) and discrete track media (DTM). For the BPM, all bits are isolated individually, while for DTM the bits are isolated in one direction only. The development of BPM is the target for HDD while the DTM development is intermediate step between the current media and BPM.

UV imprint uses template/stamp with certain flexibility, such as J-FIL, which seems to be more promising as the UV curing resist is much faster than thermal process and the template flexibility ensures a firm contact between the template and substrates. The flexibility and rigidity of the template need to be well balanced because the deformation of template reduces the patterning resolution and increases distortion of the patterned feature, which will affect track-to-track period and the down track bit-to-bit period. For 1 Tb/in^2 patterned media, the tolerable deviation of position of the bits off and down the track is about ~3nm. [52] Also resolution throughput of fabrication is another main concern. A 300 disk/h tool, the NuTera HD7000, is being used to demonstrate fesibility of a high volume J-FIL process. [53, 54] Thus the nanoimprint, like J-FIL tool strategy, has proved to be a promising option for patterned media.

Master fabrication and duplication development has also huge effect on the cost. For a 65mm DTM master at 50nm track pitch, master fabrication takes 22 days using e-beam lithography. [20] The combination of e-beam writing and directed self-assembly (DSA) is likely to be used for the fabrication of patterned media masters as shown in Figure 7. The multi-copy of templates is created by transferring the E-beam-DSA created pattern in to a proper substrate to form template.

REFERENCES

[1] G. Choe, B. R. Acharya, K. E. Johnson, K. J. Lee. *IEEE Trans. Magn.* 39(5), 2264, (2003).

[2] R. H. Victora, J. Xue, M. Patwari. IEEE Trans. Magn. 38, 1886, (2002).

[3] R. Sbiaa and S. N. Piramanayagam, *Recent Patents on Nanotechnology* 1, 29, (2007).

[4] G. A. Bertero, D. Wachenschwanz, S. Malhotra, S. Velu, B. Bian, D. Stafford, W. Yan, T. Yamashita, S. X. Wang. *IEEE Trans. Magn.* 38, 1627, (2002).

[5] T. Hikosaka, T. Komai, Y. Tanaka, *IEEE Trans. Magn.* 30, 4026, (1994).

[6] T. Oikawa, M. Nakamura, H. Uwazumi, T. Shimatsu, H. Muraoka, Y. Nakamura, *IEEE Trans. Magn.* 38, 1976, (2002).

[7] D. Ravelosona, C. Chappert, V. Mathet, H. Bernas. *Appl. Phys. Lett.*76, 236, (2000).

[8] O. Dmitrieva, B. Rellinghaus, J. Kästner, M. O. Liedke, J. Fassbender. *J. Appl. Phys.* 97, 10N1121, (2005).
[9] H. Katayama, S. Sawamura, Y. Ogimoto, J. Nakajima, K. Kojima, and K. Ohta, *J. Magn. Soc. Jpn.* 23, 233, (1999).
[10] H. Saga, H. Nemoto, H. Sukeda, and M. Takahashi, *Jpn. J. Appl. Phys.*, 38, 1839, (1999).
[11] J. Justice et al *Nature Photonics* 6, 612(2012).
[12] S. Y. Chou, *Proc. IEEE* 85, 652, (1997).
[13] C. A. Ross, *Annu. Rev. Mater. Res.* 31, 203, (2001).
[14] B. D. Terris and T. Thomson, *J. Phys. D: Appl. Phys.* 38 R199, (2005).
[15] *S.Y. Chou*, P.R. Krauss and P.J. Renstrom, *Science* 272 85, (1996).
[16] S. P. Li*et al. J. Appl. Phys.* 91, 9964, (2002).
[17] A. Lebib, S. P. Li, M. Natali *et al. J. Appl. Phys.* 89, 3892, (2001).
[18] S. Gilles, M. Meier, M. Prömpers, A. van der Hart, C. Kügeler, A. Offenhäusser, D. Mayer. Microelectronic Engineering 86, 661, (2009).
[19] M. Natali, A. Popa, U. Ebels, Y. Chen, S. P. Li and M. E. Welland, *J. Appl. Phys.* 96, 4334,(2004).
[20] Y. Chen *et al., Eur. Phys. J.* AP12, 223, (2000).
[21] M. Malloy and C. C. Litt J. Micro/nanolithography, MEMS, and MOEMS, 10, 32001, (2011).
[22] C. Peroz, S. Dhuey, M. Volger, Y. Wu, D. Olynick, and S. Cabrini, *Nanotechnology* 21, 445301, (2010).
[23] T. Thum-Albrecht et al. *Science* 290, 2126, (2000).
[24] T. Ghosal, T. Maity, J. F. Godsell, S Roy, M A Morris; *Adv. Mater.*, 24, 2390, (2012).
[25] Ion Bita*et al. Science* 321, 939, (2008).
[26] J. Y. Cheng et al. *Adv. Mater* 20, 3255, (2008).
[27] R. Ruiz et al. *Science* 321, 936, (2008).
[28] M.Hehn, K.Ounadjela, J. P. Bucher, F.Rousseaux, D.Decanini, B. Bartenlian and C.Chappert, *Science* 272, 1782, (1996).
[29] E. Gu et al. Phys. Rev. Lett. 78, 1158, (1997).
[30] R. P. Cowburn, D. K. Koltsov, A. O. Adeyeye, M. E. Welland, and D. M. Tricker, *Phys. Rev. Lett.*83 1042, (1999).
[31] M. Natali et al. *Phys. Rev. Lett.*, 88, 157203, (2002).
[32] A. Lebib et al. *J. Appl. Phys.* 89, 3892, (2001).
[33] S. P. Li et al. *Phys. Rev. Lett.* 86, 1102, (2001).
[34] Y. B. Xu et al. *J. Appl. Phys.* 87, 7019, (2000).
[35] R. P. Cowburnand M. E. Welland *Science* 287, 1466, (2000).

[36] D. A. Allwood, G. Xiong, C. C. Faulkner, D. Atkinson, D. Petit, and R. P. Cowburn, *Science* 309, 1688, (2005).
[37] J. Gadbois and J-G Zhu, *IEEE Trans Magn.* 31, 3802, (1995).
[38] S. Jain, A. O. Adeyeye, and N. Singh, *Nanotechnology* 21, 1, (2010).
[39] T. Shinjo et al. *Science* 289, 930, (2000).
[40] O. Fruchart et al. *Phys. Rev.* B 70, 172409, (2004).
[41] E. C. Stoner and E. P. Wohlfarth, *Philos. Trans. R. Soc. London* A 240, 74, (1948).
[42] C. Tannous and J. Gieraltowski, *Eur. J. Phys.* 29, 475, (2008).
[43] C. Thirion et al. *J. Magn. Mag. Mat.* 993–995, 242, (2002).
[44] C. Chappert et al. *Science* 280, 1919, (1998).
[45] I. L. Prejbeanu et al. *J. Appl. Phys.* 91, 7343, (2002).
[46] S. Evoy, D. W. Carr, L. Sekaric, Y. Suzuki, J. M. Parpia, and H. G. Craighead, *J. Appl. Phys.* 87, 404, (2000).
[47] C. Haginoya et al. *J. Appl. Phys.* 85, 8327, (1999).
[48] R. Rottmayer, C. Cheng, X. Shi, L. Tong, and H. Tong, *US Patent* 5986978, (1999).
[49] L. Yin et al., *Appl. Phys.Lett.,* vol. 85, no. 3, 467, (2004).
[50] X. Shi and J. Hesselink, *J. Appl. Phys.,* vol. 41, 1632, (2003).
[51] R. Grober, S. Bukofsky, and S. Seeberg, *Appl. Phys. Lett.,* vol. 70, 2368, (1997).
[52] E. A. Dobisz, *proceedings of the IEEE* 96, 1836,(2008).
[53] L. Singh et al. *Proc. SPIE* 7970, 797007, (2011).
[54] M. Malloy et al. *Proc. SPIE* 7970, 797006, (2011).

In: Research Advances in Magnetic Materials ISBN: 978-1-62417-913-6
Editors: C. Toulson and D. Marwick

Chapter 3

PHOTOMAGNETIC ORGANIC-INORGANIC HYBRID MATERIALS

***Masashi Okubo*[*1] *and Norimichi Kojima*[2]**
[1]National Institute of Advanced Industrial Science and Technology, Tsukuba, Ibaraki, Japan
[2]Graduate School of Arts and Sciences, The University of Tokyo, Meguro-ku, Tokyo, Japan

ABSTRACT

The development of new magnetic materials is a key challenge in recent chemistry and physics. The design of magnetic coordination polymers has attracted much attention, because the intrinsic tunability of both the electronic and structural properties provides potential for multifunctional magnets. Therefore, the magnetic coordination polymers controllable by the external stimuli, which can be applied to electronic switching devices, have been investigated intensively. In particular, photomagnetism, *i.e.*, controllable magnetism by light irradiation, is one of the best studied multifunctional magnetisms. In this chapter, the recent progress in the development of photomagnetic organic-inorganic hybrid materials is summarized.

[*] To whom correspondence should be addressed: m-okubo@aist.go.jp, cnori@ecc.mail.u-tokyo.ac.jp, National Institute of Advanced Industrial Science and Technology, Umezono 1-1-1, Tsukuba, Ibaraki 305-8568, Japan.

1. PHOTOMAGNETISM

Switching magnetism is a magnetic property which can be controlled by the external stimuli such as temperature, pressure, light or humidity. In particular, photomagnetism, *i.e.*, controllable magnetism by light irradiation, is one of the best studied switching magnetism, because of industrial demands for the application to electronic devices.

In general, a photomagnet should have a metastable state. When bistable magnetic states can be switched via light excited states, photomagnetism is achieved (Figure 1). For example, transition metal complexes with d^4 to d^7 electron configuration can exist either in the low-spin (LS) state or high spin (HS) state. When the LS state is the ground state at low temperature, a spin transition to the HS state can occur by increasing temperature.

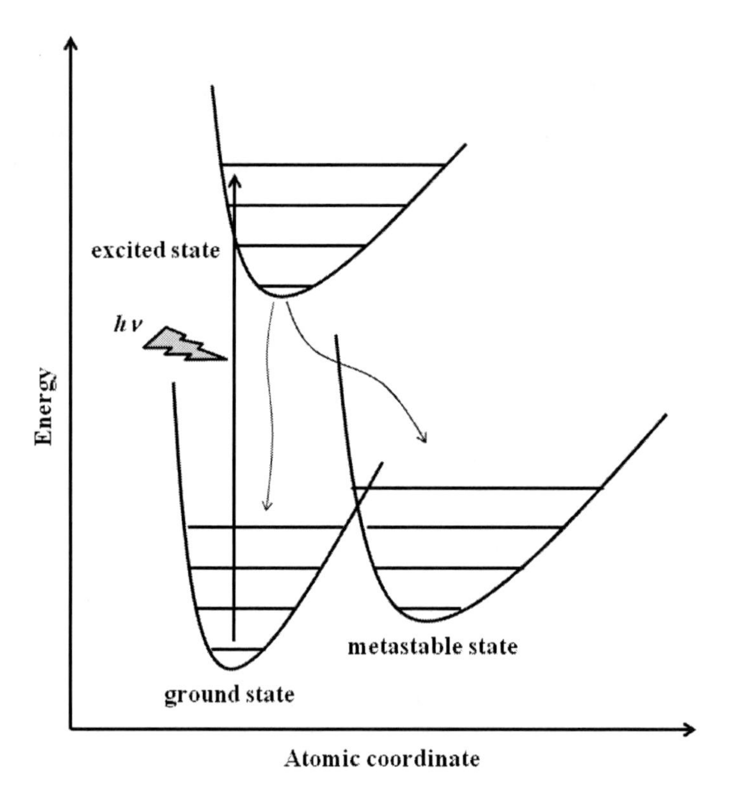

Figure 1. Potential curves for the ground, light-excited and metastable states in a photomagnet.

The spin transition between the HS and LS states is generally called as spin crossover (SC).[1] The SC transition was first discovered in 1931.[2] Whether the SC transition is observable at the accessible temperature depends on the energy gap between the e_g and t_{2g} orbitals and the mean spin pairing energy, however, the existence of bistable magnetic states allows the occurrence of photomagnetism, *i.e.*, light-induced excited spin state trapping (LIESST).[3] A typical LIESST complex $[Fe^{II}(ptz)_6](BF_4)_2$ (ptz: 1-propyltetrazole) shows SC from the low spin state (t_{2g}^6, $S = 0$) to the metastable high spin state ($t_{2g}^4e_g^2$, $S = 2$) by light irradiation corresponding to the spin allowed d-d transition in the low spin state.[4]

The bistable magnetic states in the mono-nuclear complex have also been discovered for valence tautomeric complexes, in which the ligand-to-metal charge transfer (LMCT) can generate a metastable magnetic state $d^{n+1}\underline{L}$ from the electronic configuration of d^n. Here, $\underline{L}$ denotes the ligand hole. The bistability through LMCT could provide the photomagnetism. For example, $[Co^{III}$(3,5-dbcat)(3,5-dbsq)(tmeda)] (3,5-dbcat:3,5-di-*tert*-butyl-1,2-catecholate, 3,5-dbsq: 3,5-di-*tert*-butyl-1,2-semiquinonate, tmeda: *N,N,N',N'*-tetramethylethylethylenediamine) shows photomagnetism by LMCT. The electron of the catecholate ligand is transferred to Co^{III} by light irradiation of 532 nm to give the metastable state of $[Co^{II}(3,5\text{-dbsq})_2(\text{tmeda})]$.[5]

In the cluster complexes (dinuclear, trinuclear, *etc.*), the inter-valence charge transfer (IVCT) is also accounted for generation of the metastable state, in addition to SC and LMCT. A magnetic ground state M_1^{n+}-L-M_2^{m+} (M_1, M_2: metal center, L: ligand) is excited to a metastable state $M_1^{(n+1)+}$-L-$M_2^{(m-1)+}$ by IVCT. A typical IVCT cluster is $[Co_2Fe_2(CN)_6(tp^*)(dtbbpy)_4](PF_6)_2\cdot 2CH_3OH$ (tp*: hydrotris(3,5-dimethylpyrazol-1-yl)borate, dtbbpy: 4,4'-di-*tert*-butyl-2,2'-bipyridine). The ground state of this cluster is $[Co^{III}_2Fe^{II}_2]$ below 260 K, but light irradiation of 808 nm induces IVCT of $Fe^{II} \rightarrow Co^{III}$ to give the metastable $[Co^{II}_2Fe^{III}_2]$.[6]

Polynuclear coordination polymers can show the bistable magnetic states through SC, LMCT, and IVCT, which results in photomagnetism. For example, Fe^{II} in $Fe^{II}_2[Nb^{IV}(CN)_8]\cdot(4\text{-pyridinealdoxime})_8\cdot 2H_2O$ shows SC from the diamagnetic Fe^{II} low spin state to the paramagnetic Fe^{II} high spin state by light irradiation of 473 nm, which results in the switching magnetism between a stable paramagnet and a metastable ferromagnet.[7]

As described above, the design of the appropriate electronic structures for SC, LMCT or IVCT is efficient to achieve photomagnetism. In contrast, fabrication of an organic-inorganic hybrid material with a photochromic organic molecule and a magnetic inorganic compound is another efficient way to design a photomagnet. In particular, layered magnetic compounds can exhibit intercalation chemistry of various ionic species, thus provide an excellent opportunity to control the magnetic properties by the intercalated photochromic molecules. [8] In this chapter, photomagnetism of some organic-inorganic hybrid materials is described.

2. Photomagnetism of Layered Hydroxides

A series of layered double hydroxides (LDHs), $M(OH)_{2-x}A_y{\bullet}zH_2O$ (M: Co, Cu, A: anion), consists of 2D triangular arrays of M^{II} ions, which are octahedrally coordinated by oxygen atoms belonging to either hydroxide or anions. Hereafter, LDHs with M = Co and Cu are denoted as Co-LDH and Cu-LDH, respectively. The 2D array is schematically illustrated in Figure 2.

In general, all the Cu-ions in Cu-LDHs are octahedrally coordinated as shown in Figure 2. On the other hand, the crystal structure of Co-LDHs is characterized into two kinds of layer structure, *i.e.*, single decker layer structure only with octahedral Co-ions and triple decker layer structure with both octahedral and tetrahedoral Co-ions. [9] The colors of the octahedral and tetrahedral Co-ions are pale pink and blue, respectively, thus Co-LDHs with the single-decker layers and triple-decker layers display pale pink and green, respectively.

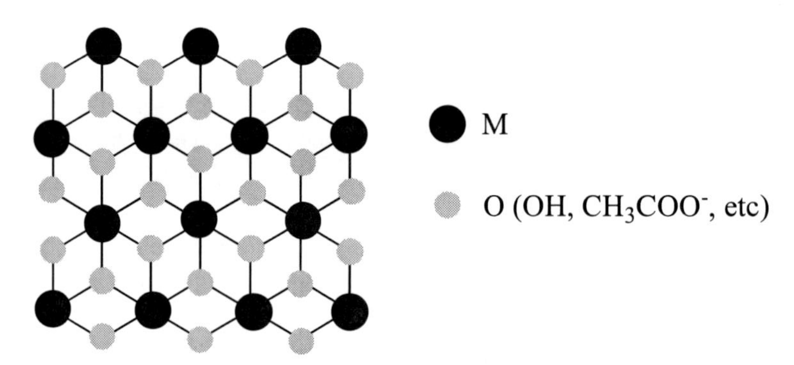

Figure 2. Schematic illustration of the 2D array of LDH.

The interlayer distance, d, of LDHs depends on the size of the intercalated anion, and can be controlled in the range of 9 to 40 Å. [10] When the long-chain carboxylate anion, n-$C_mH_{2m+1}CO_2^-$, are intercalated into Cu-LDHs, the interlayer distance varies linearly with the length of the n-alkyl chain as $d = d_0$

+ $1.27\eta m\cos\theta$, where η is the number of layers of aliphatic chains ($\eta = 1$ for mono-layer packing and 2 for bi-layer packing), and θ is the tilt angle of the chains.[10] The packing of the aliphatic chains essentially depends on the water content; the LDH with mono-layer packing (α-phase) is hydrated while one with bi-layer packing (β-phase) is anhydrous. Note that d_0 is the sum of the size of the bridging group, the van der Waals distance between terminal methyl groups, and the thickness of the inorganic layer.

The magnetic properties of LDHs strongly depend on the intercalated anions. In case of Cu-LDHs, the $m = 0$ and 1 compounds show the meta-magnetism, while the $m = 7$ to 9 compounds are spin canted weak ferromagnets below 22 K. [11] The intralayer interaction between Cu-ions changes from ferromagnetic ($m = 0$ and 1) to antiferromagnetic ($m = 7$, 8 and 9), because the Cu-OH-Cu exchange interaction is sensitive to the Cu-OH-Cu angle. [12] For example, di-□-hydroxo Cu dimers, $[Cu(L)(OH)]_2^{2-}$ (L: bidentate nitrogen-containing ligand), show the variation of the Cu-OH-Cu exchange interaction from +172 cm^{-1} (Cu-OH-Cu angle of 95.5°) to 509 cm^{-1} (Cu-OH-Cu angle of 104.1°) due to the angle-sensitive kinetic exchange (antiferromagnetic contribution) and the angle-insensitive potential exchange (ferromagnetic contribution). [3] The drastic change of the intralayer interaction for Cu-LDHs can be ascribed to the change of the Cu-OH-Cu angle.

It should be emphasized that Cu-LDHs with a large interlayer distance exhibit the ferromagnetic transition at rather high temperature, although the layered systems with isotropic intralayer interaction cannot show long range order at finite temperature. The intralayer anisotropic exchange interaction may be one of the origins for the ferromagnetic transition. Furthermore, since the interlayer dipolar interaction depends on the square of the effective moment, the high spin state of the short range ordered domain in the ferromagnetic layer can make the interlayer dipolar interaction efficient enough to show the long range order at rather high temperature. [11]

Figure 3 shows χT as a function of temperature for a typical pale pink Co-LDH with single-decker layers, $Co_2(OH)_3(CH_3COO)\cdot H_2O$. [13] The octahedral Co^{II} in Co-LDHs has an electron configuration of $t_{2g}^5 e_g^2$ ($S = 3/2$, $L = 1$), thus due to the spin-orbit coupling, the drop in the χT value occurs with decreasing temperature down to 60 K. Note that a green Co-LDH with triple-decker layers also shows the drop in the χT value with decreasing temperature. [14] As for a green Co-LDH with triple-decker layers, the temperature dependence of χT is explained by both the spin-orbit coupling of Co-ion and the antiferromagnetic interaction between the tetrahedral and octahedral Co-ions.

Co-LDHs with the *n*-alkyl chain anions show the long range magnetic order (ferromagnetic/ferrimagnetic transition) due to the intralayer anisotropic exchange interaction and interlayer dipolar interaction. As shown in Figure 4, the Curie temperature, T_C, of Co-LDHs with n-$C_mH_{2m+1}CO_2^-$ decreases from 14 K to 10 K with increasing the inter-layer distance, *d*, from 13 Å ($m = 1$) to 25 Å ($m = 12$). [10].

The interlayer interaction in LDHs can be enhanced by introduce the π–electron system. When the C_6 dianions, *i.e.*, hexanedioate ($C_6H_{12}(CO_2^-)_2$) without π–electron and (*E*,*E*)-2,4-hexadienedioate ($C_6H_8(CO_2^-)_2$) with π-electron, are intercalated into Co-LDHs, *d* is almost same for two compounds (14.7 Å). However, T_C is 26 K for Co-LDH with (*E*,*E*)-2,4-hexadienedioate, which is much higher than that for Co-LDH with hexanedioate (Figure 5). [14] The exchange interaction via delocalized π-electrons, which is stronger than the dipolar interaction, should contribute to the enhanced T_C for Co-LDH with (*E*,*E*)-2,4-hexadienedioate.

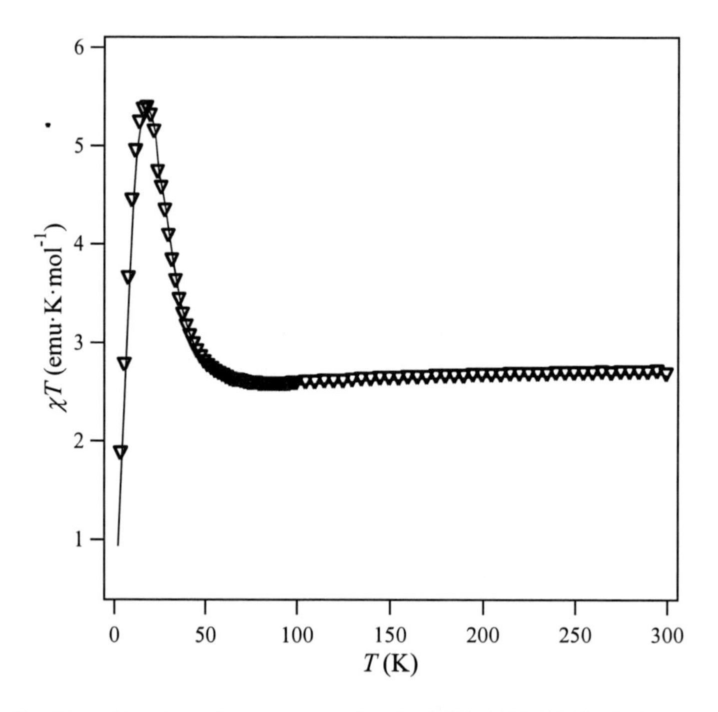

Figure 3. χT as a function of temperature for $Co_2(OH)_3(CH_3COO)\cdot H_2O$.

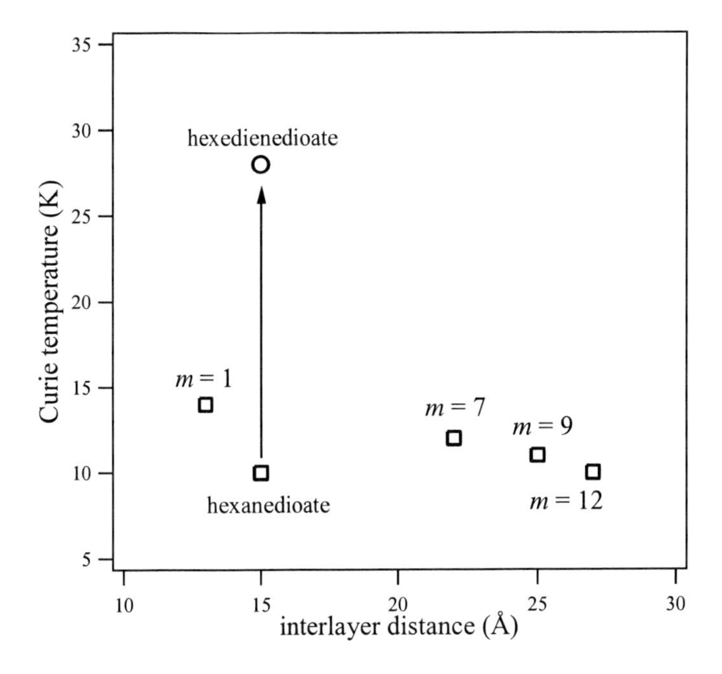

Figure 4. The Curie temperatures of Co-LDHs with n-$C_mH_{2m+1}CO_2^-$ (m = 1, 7, 9, 12), hexanedioate and hexediendioate as a function of interlayer distance d. The arrow shows the enhancement of Curie temperature induced by the delocalized π electron system of intercalated anion.

Figure 5. Photochromic reaction of diarylethene. *1a*: open form with R = SO_3^-, *1b*: closed form with R = SO_3^-.

Summarizing the magnetic properties of LDHs,

1. Intralayer interaction is sensitive to the intercalated anions.
2. Interlayer interaction depends on d when the dipolar interaction is dominant.

3. Interlayer interaction is significantly enhanced when the intercalated anion has the conjugated π-electron system.

Therefore, when the photochromic anion is intercalated into LDHs, the photo-isomerization is expected to affect both the intra-layer and interlayer interactions, resulting in photomagnetism. Furthermore, if the photo-isomerization accompanies the change in the π-electron system, the interlayer interaction could be controlled more drastically. Thus, a photochromic diarylethene anion is intercalated into Co-LDH to form a photomagnetic organic-inorganic hybrid material.

Diarylethenes undergo a thermally irreversible and fatigue resistant photochromic reaction between the open and closed form (Figure 5). [15] According to the molecular orbital calculation, the π-electron system is localized in two units for the open form, while the π-electron is delocalized within the entire molecule for the closed form.[16] Therefore, intercalation of a diarylethene anion (Figure 5, **1a** or **1b**) as a magnetic coupler into Co-LDH is expected to provide a photomagnet.

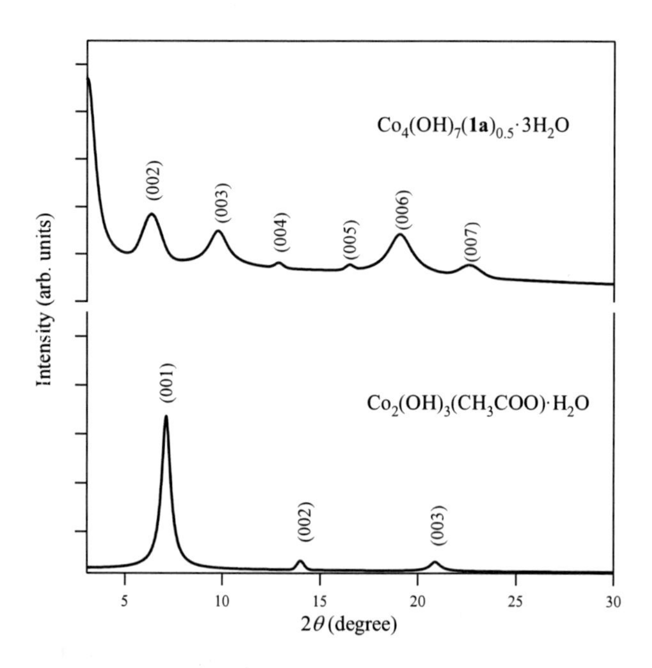

Figure 6. X-ray diffraction patterns for $Co_2(OH)_3(CH_3COO)\cdot 3H_2O$ and $Co_4(OH)_7(1a)_{0.5}\cdot 3H_2O$.

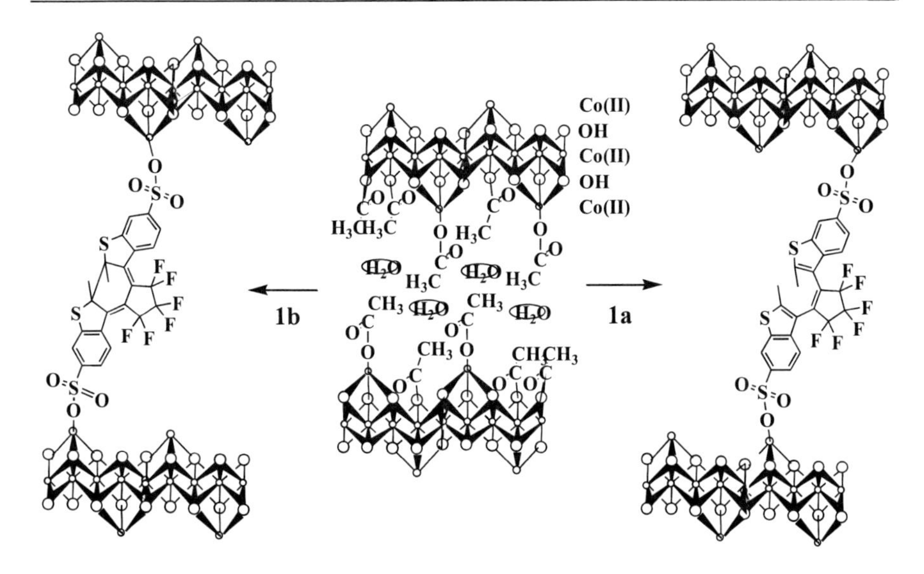

Figure 7. Schematic illustration of anion exchange reaction and structures of Co-LDHs with intercalated open/closed diarylethenes.

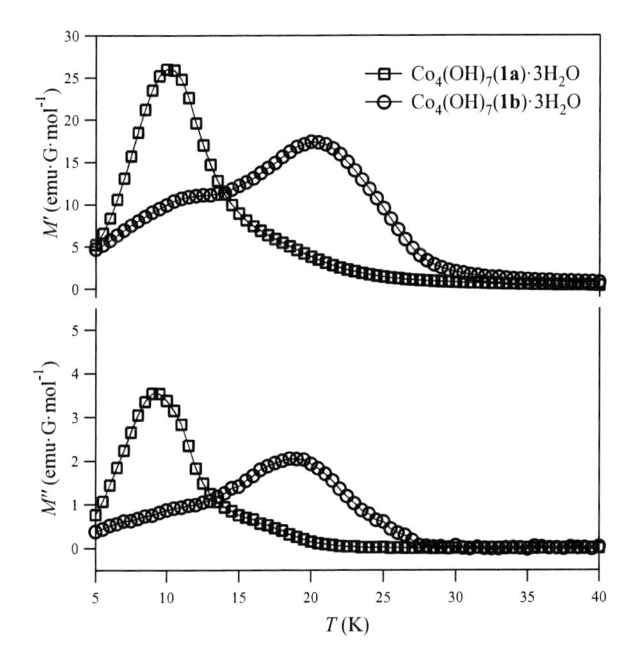

Figure 8. Real (M') and imaginary (M'') part of ac magnetic susceptibility for $Co_4(OH)_7(1a)_{0.5}\cdot 3H_2O$ and $Co_4(OH)_7(1b)_{0.5}\cdot 3H_2O$.

The intercalation compound $Co_4(OH)_7(\mathit{1a}$ or $\mathit{1b})_{0.5}\cdot 3H_2O$ can be obtained by an anion-exchange reaction from $Co_2(OH)_3(CH_3COO)\cdot 3H_2O$. Figure 6 shows the XRD patterns for $Co_2(OH)_3(CH_3COO)\cdot 3H_2O$ and $Co_4(OH)_7(\mathit{1a})_{0.5}\cdot 3H_2O$. [13] All the intense (00*l*) diffractions shifted, which suggests that the anion exchange occurred homogeneously. The calculated *d* is expanded from 12.8 Å to 27.8 Å by the intercalation of *1a*. The schematic illustration of the intercalated compounds is shown in Figure 7. The photoisomerization of *1a* in $Co_4(OH)_7(\mathit{1a})_{0.5}\cdot 3H_2O$ occurs reversibly upon light irradiation of 313 nm (*1a* to *1b*) and 550 nm (*1b* to *1a*). [13]

The ac magnetic susceptibility for $Co_4(OH)_7(\mathit{1a})_{0.5}\cdot 3H_2O$ and $Co_4(OH)_7(\mathit{1b})_{0.5}\cdot 3H_2O$ showed an abrupt increase at 9 and 20 K, respectively (Figure 8). [14] The LDH intercalated with the closed form of the diarylethene showed higher T_C than that with the open form. Both the intra-layer interaction and inter-layer dipolar interaction should be same regardless of the open/closed forms, thus the interlayer exchange interaction may be reinforced via delocalized π-electron system of the closed form, which could result in higher T_C.

The correlation between the interlayer interaction and T_C can be described theoretically. [17] The magnetic interaction in Co-LDHs is expressed by the Hamiltonian for a quasi-2D Ising spin system with the interlayer interaction as,

$$H = -2J\sum_{\text{intralayer}}(S_i^z S_j^z)\ -2J\xi\sum_{\text{interlayer}}(S_i^z S_j^z)$$

where J is the intralayer exchange interaction, S_i^z is Co^{II} Ising spin, and $0 \leq \xi \leq 1$. The system with $\xi = 0$ represents a 2D Ising spin system, while the system with $\xi = 1$ represents a 3D Ising spin system. The Monte Carlo calculation with the above Hamiltonian gives the ξ dependence of T_C, in which the increase in the interlayer exchange interaction enhances T_C. The theoretical ratio of $T_C(\xi{=}1)/T_C(\xi{=}0)$ is calculated as 2.0.

As for the Co-LDHs intercalated with the closed/open forms, the ratio of T_C is 2.2, which is close to the theoretical ratio of T_C for the 2D- and 3D-limits. Thus, $Co_4(OH)_7(\mathit{1a})_{0.5}\cdot 3H_2O$ can be regarded as a quasi-2D magnet (2D Ising spin layer with weak interlayer dipolar interaction), while $Co_4(OH)_7(\mathit{1b})_{0.5}\cdot 3H_2O$ can be regarded as a quasi-3D magnet (2D Ising spin layer with strong interlayer exchange interaction). Therefore, photoisomerization of diarylethene successfully achieved the switching magnetism between the quasi-2D and quasi-3D magnets.

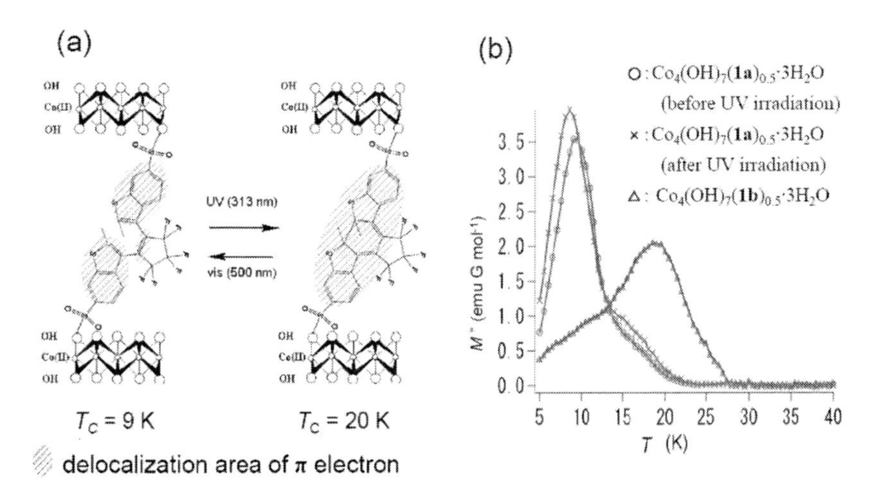

Figure 9. (a) Schematic representation of the photo-induced conversion between $Co_4(OH)_7(1a)_{0.5}\cdot 3H_2O$ and $Co_4(OH)_7(1b)_{0.5}\cdot 3H_2O$. (b) Imaginary (M'') part of the ac magnetic susceptibility as a function of temperature for $Co_4(OH)_7(1a)_{0.5}\cdot 3H_2O$ before and after UV irradiation at room temperature, and $Co_4(OH)_7(1b)_{0.5}\cdot 3H_2O$.

Figure 9 shows the ac magnetic susceptibilities as a function of temperature for $Co_4(OH)_7(1a)_{0.5}\bullet 3H_2O$ before and after UV irradiation of 313 nm at room temperature and for $Co_4(OH)_7(1b)_{0.5}\bullet 3H_2O$. [14] As shown in Figure 9, after UV irradiation of 313 nm, small fraction of a new phase transition at 18 K was observed. Comparing this shoulder and the ac magnetic susceptibility for $Co_4(OH)_7(1b)_{0.5}\bullet 3H_2O$, the photo-induced phase transition around 18 K for $Co_4(OH)_7(1a)_{0.5}\bullet 3H_2O$ should correspond to the ferromagnetic phase transition which is observed in $Co_4(OH)_7(1b)_{0.5}\bullet 3H_2O$. Generally speaking, when the UV light is irradiated on diarylethene, the light could not penetrate into the inner sites of the sample because of the strong optical absorption corresponding to the $\pi - \pi^*$ transition of diarylethene. The photo-isomerization of all the diarylethene molecule from the open form to the close one could not take place, especially at the inner sites. Therefore, the UV irradiated $Co_4(OH)_7(1a)_{0.5}\bullet 3H_2O$ should have two domains corresponding to $Co_4(OH)_7(1a)_{0.5}\bullet 3H_2O$ and $Co_4(OH)_7(1b)_{0.5}\bullet 3H_2O$. The two magnetic phase transitions observed for the UV irradiated $Co_4(OH)_7(1a)_{0.5}\bullet 3H_2O$ can be explained in this manner. In fact, as shown in Figure 9, the conversion ratio of diarylethene from the open form to the close one in $Co_4(OH)_7(1a)_{0.5}\bullet 3H_2O$ is estimated at less than 5 % based on the ac magnetic susceptibility. In order to avoid the $\pi-\pi^*$ absorption of diarylethene and produce a remarkable photo-conversion from $Co_4(OH)_7(1a)_{0.5}\bullet 3H_2O$ to $Co_4(OH)_7(1b)_{0.5}\bullet 3H_2O$, the photo-isomerization of diarylethene through two-photon excitation by 630 nm light would be necessary.

3. Photomagnetism of Dithiooxalate-Bridged Compounds

In the case of mixed-valence systems whose spin states are situated in the spin-crossover region, it is expected that new types of conjugated phenomena coupled with spin and charge take place between different metal ions in order to minimize the free energy in the whole system. Based on this viewpoint, an organic-inorganic hybrid system, $(n\text{-}C_nH_{2n+1})_4N[Fe^{II}Fe^{III}(dto)_3]$(dto = $C_2O_2S_2$) has been synthesized, and a new type of phase transition called charge transfer phase transition was discovered in 2001. [18, 19] In this system, the Fe^{II} and Fe^{III} atoms are alternately bridged by dithiooxalato molecules, which forms the 2D honeycomb network structure of $[Fe^{II}Fe^{III}(dto)_3]$. [20] The $(n\text{-}C_3H_7)_4N$ cation layer is intercalated between two adjacent $[Fe^{II}Fe^{III}(dto)_3]$ layers, which is shown in Figure 10. Fe^{II} and Fe^{III} sites in $(n\text{-}C_3H_7)_4N[Fe^{II}Fe^{III}(dto)_3]$ are coordinated by six O and six S atoms at room temperature, respectively.

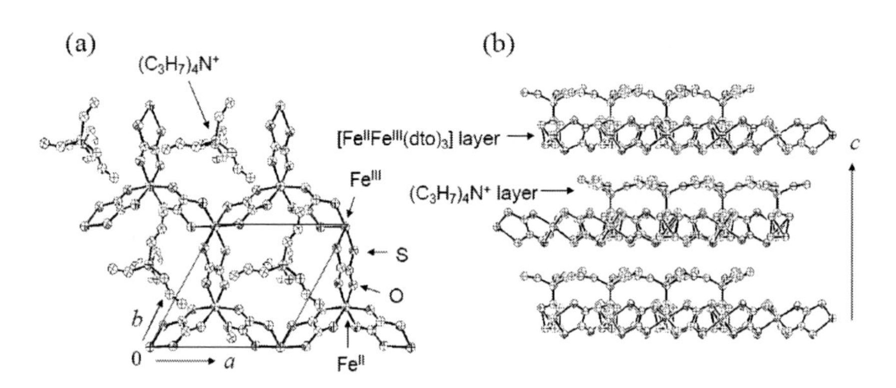

Figure 10. Crystal structure of $(n\text{-}C_3H_7)_4N[Fe^{II}Fe^{III}(dto)_3]$ at room temperature. (a) [001] projection, (b) [210] projection.

Figure 11 shows the ^{57}Fe Mössbauer spectra for $(n\text{-}C_nH_{2n+1})_4N[Fe^{II}Fe^{III}(dto)_3]$ (n = 3 - 5). The assignment of the spectra A, B, C and D were confirmed by the ^{57}Fe Mössbauer spectra of $(n\text{-}C_3H_7)_4N[^{57}Fe^{II}Fe^{III}(dto)_3]$ and $(n\text{-}C_3H_7)_4N[Fe^{II\,57}Fe^{III}(dto)_3]$. [21] At 200 K, the line profiles of all the complexes are quite similar to each other. The isomer shift (*IS*) and the quadrupole splitting (*QS*) of the spectrum A at 200 K for (n-$C_nH_{2n+1})_4N[Fe^{II}Fe^{III}(dto)_3]$ (n = 3 - 5) are quite similar to those (*IS* = 0.33 mm/s, *QS* = 0.35 mm/s at 196 K) of the ^{57}Fe Mössbauer spectrum for the Fe^{III}(S = 1/2) site in $KBa[Fe^{III}(dto)_3]\cdot 3H_2O$, [22] where the Fe^{III} site is coordinated by six S atoms. On the other hand, the *IS* and *QS* of the spectrum B at 200 K are quite similar to those (*IS* = 1.235 mm/s, *QS* = 1.42 mm/s at 190 K) of the ^{57}Fe Mössbauer spectrum for the Fe^{II}(S = 2) site in (n-$C_4H_9)_4N[Fe^{II}Fe^{III}(ox)_3]$(ox = oxalato),[23] where the Fe^{II} site is coordinated by six O atoms. Therefore, it is concluded that the Fe^{II}(S = 2) and Fe^{III}(S = 1/2) sites in (n-$C_nH_{2n+1})_4N[Fe^{II}Fe^{III}(dto)_3]$(n = 3 - 5) are coordinated by six O atoms and six S atoms, respectively.

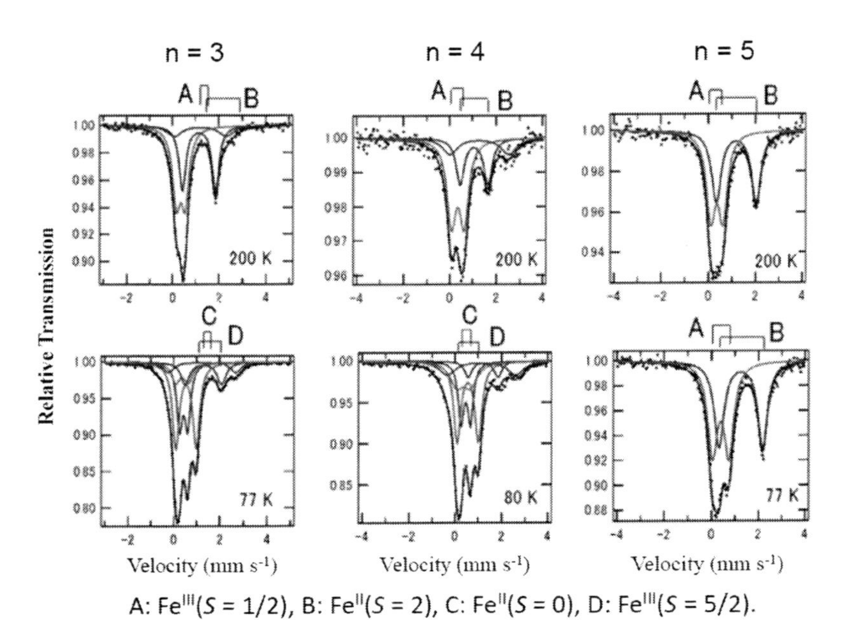

Figure 11. ^{57}Fe Mössbauer spectra for (n-$C_nH_{2n+1})_4N[Fe^{II}Fe^{III}(dto)_3]$(n = 3 – 5). A: Fe^{III}(S = 1/2), B: Fe^{II}(S = 2), C: Fe^{II}(S = 0), D: Fe^{III}(S = 5/2).

In the cases of n = 3 and 4, at 77 K (80 K for n = 4), the spectra, A and B, decrease by about 80%. Instead of these spectra, the spectra, C and D, appear. The *IS* and *QS* of the spectrum, C, for n = 3 and 4 are similar to those (*IS* = 0.325 mm/s, *QS* = 0.39 mm/s) of the Fe^{II}(S = 0) site in $[Fe^{II}(bipy)_3](ClO_4)_2$ (bipy = 2,2'-bipyridine). [24] On the other hand, the *IS* and QS of the spectrum, D, for n = 3 and 4 are similar to those (*IS* = 0.486, *QS* = 0.64 at 90

K) of the ^{57}Fe Mössbauer spectrum for the $Fe^{III}(S = 5/2)$ site in $(n\text{-}C_4H_9)_4N[Fe^{II}Fe^{III}(ox)_3]$. [23]

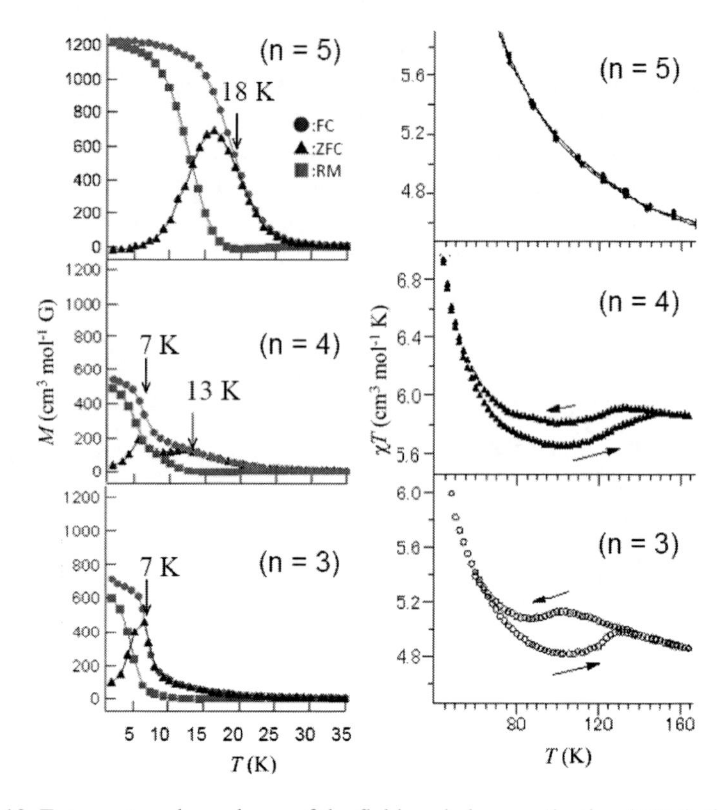

Figure 12. Temperature dependence of the field cooled magnetization (FCM), the remnant magnetization (RM), the zero-field cooled magnetization (ZFCM), and χT for $(n\text{-}C_nH_{2n+1})_4N[Fe^{II}Fe^{III}(dto)_3]$ (n = 3 – 5). In the measurement of FCM and ZFCM, the applied magnetic field was 30 Oe. In the case of χT, the applied magnetic field was 5000 Oe.

From the ^{57}Fe Mössbauer spectra for $(n\text{-}C_nH_{2n+1})_4N[Fe^{II}Fe^{III}(dto)_3]$ (n = 3 and 4), it is obvious that the charge transfer phase transition (CTPT) takes place between 200 K and 77 K for n = 3 and 4. The coexistence of the higher and lower temperature phases at 77 K is typical of first order phase transition, which reflects on the thermal hysteresis in magnetic susceptibility as shown in Figure 12. [25] From the analysis of heat capacity, the critical temperature of the CTPT was determined at 122.4 K for n = 3. [26] In this way, $(n\text{-}C_nH_{2n+1})_4N[Fe^{II}Fe^{III}(dto)_3]$ (n = 3 and 4) undergo a thermally induced CTPT at about 120 K where the electron transfer occurs reversibly between the t_{2g}

orbitals of the Fe^{II} and Fe^{III} sites, which is schematically shown in Figure 13. In the case of n = 5, however, the line profile of the ^{57}Fe Mössbauer spectra remains unchanged between 200 K and 77 K, which implies that the CTPT does not take place for n = 5. In fact, the higher temperature phase exists between 300 K and 4 K for n = 5.

All of $(n\text{-}C_nH_{2n+1})_4N[Fe^{II}Fe^{III}(dto)_3]$ (n = 3 - 5) undergo ferromagnetic transitions, [25] which is shown in Figure 12. In the case of n = 3, the low temperature phase (LTP) with $Fe^{II}S_6$ ($S = 0$) and $Fe^{III}O_6$ ($S = 5/2$) undergoes the ferromagnetic transition at T_C = 7 K, while, in the case of n = 5, the high temperature phase (HTP) with $Fe^{II}O_6$ ($S = 5/2$) and $Fe^{III}S_6$ ($S = 1/2$) undergoes the ferromagnetic transition at T_C = 23 K. In the case of n = 4, the LTP and HTP coexist even at low temperature and these two phases undergo the ferromagnetic transitions of T_C = 7 K and 13 K, respectively.

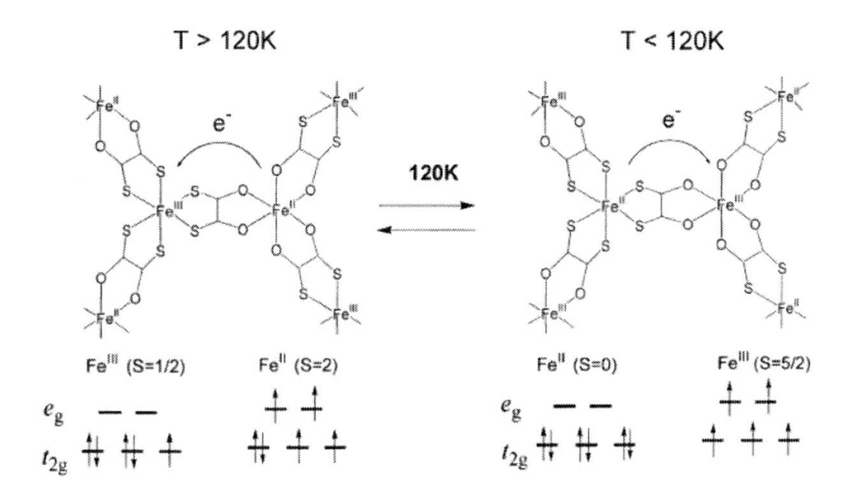

Figure 13. Schematic feature of the charge transfer phase transition in $(n\text{-}C_3H_7)_4N[Fe^{II}Fe^{III}(dto)_3]$.

Now, let us consider the mechanism of CTPT. The driving force responsible for the CTPT is considered to be the difference in the spin entropy between the HTP and LTP. It should be noted that the spin entropy in the HTP is $R\ln(2\times5)$ = 19.15 $JK^{-1}mol^{-1}$ and that in the LTP is $R\ln(1\times6)$ = 14.90 $JK^{-1}mol^{-1}$, where R is the gas constant. Therefore, the spin-entropy gain expected from the CTPT is estimated at 4.25 $JK^{-1}mol^{-1}$. Since the observed entropy gain at the CTPT in n = 3 is 9.20 $JK^{-1}mol^{-1}$, [26] the entropy change originating in intra-molecular vibration is quite smaller than that in normal spin-crossover transition. For example, about 35 $JK^{-1}mol^{-1}$ was estimated for the vibrational contribution to the entropy change in the spin-crossover phenomenon observed in $[Fe(phen)_2(NCS)_2]$. [27] The CTPT strongly depends on applied pressure. In the case of n = 3, when a hydrostatic pressure is applied up to 1.0 GPa, the transition temperature of the CTPT linearly increases from 120 K to 220 K. [28] On the other hand, $(n\text{-}C_nH_{2n+1})_4N[Fe^{II}Fe^{III}(dto)_3]$ (n = 5) does not show the CTPT between 2 K and 300 K at ambient pressure. However, when a hydrostatic pressure is applied up to 0.5 GPa for n = 5, the CTPT appears suddenly at about 100 K, then the transition temperature increases linearly with increasing applied pressure. [28]

Taking account of the behavior of CTPT under various applied pressures, the phase diagram of $(n\text{-}C_nH_{2n+1})_4N[Fe^{II}Fe^{III}(dto)_3]$ is schematically shown in Figure 14, where we assume that the difference of enthalpy (ΔH) between LTP and HTP does not depend on temperature. The vertical axis denotes the difference of Gibbs energy ($G = H - TS$) between HTP and LTP. The slope of the curve corresponding to the HTP is the difference of spin entropy (ΔS = 4.25 $JK^{-1}mol^{-1}$) between HTP and LTP. T_{CT} is the phase transition temperature of CTPT. In the case of n = 3, as shown in Figure 14(a), the Gibbs energy of LTP is lower than that of HTP at $T = 0$, consequently the CTPT takes place at a finite temperature. The transition temperature of CTPT linearly increases with increasing applied pressure, which implies that the applied pressure increases the Gibbs energy of HTP relative to that of LTP. On the other hand, in the case of n = 5, the Gibbs energy of HTP is lower than that of LTP at $T = 0$, consequently the CTPT does not take place at finite temperature. As shown in Figure 14(b), when an external pressure is applied, the applied pressure increases the enthalpy of HTP relative to that of LTP and ΔH becomes to be small, and eventually the crossover of the Gbbs energies for LTP and HTP occurs at a finite temperature, which implies the pressure-induced CTPT for n = 5.

As mentioned above, the CTPT and the ferromagnetic transition in $(n\text{-}C_nH_{2n+1})_4N[Fe^{II}Fe^{III}(dto)_3]$ remarkably depend on the size of intercalated cation, which implies a possibility to control the CTPT and the ferromagnetism in the $[Fe^{II}Fe^{III}(dto)_3]^-_\infty$ layers by means of the isomerization of intercalated cation. In organic-inorganic hybrid systems, it is effective to use an organic photochromic molecule for producing photo-switchable materials. On the basis of this strategy, (SP-R)$[Fe^{II}Fe^{III}(dto)_3]$ (SP-R = cationic spiropyran) has been synthesized.[29] In general, the cationic spiropyran is converted from the yellow-colored closed form (CF) to the red-colored open form (OF) upon the irradiation of UV light (330-370 nm). The OF is usually less stable and returns to the CF both thermally and photochemically (500 - 600 nm). This photo-isomerization is associated with the large volume change, which is schematically shown in Figure 15.

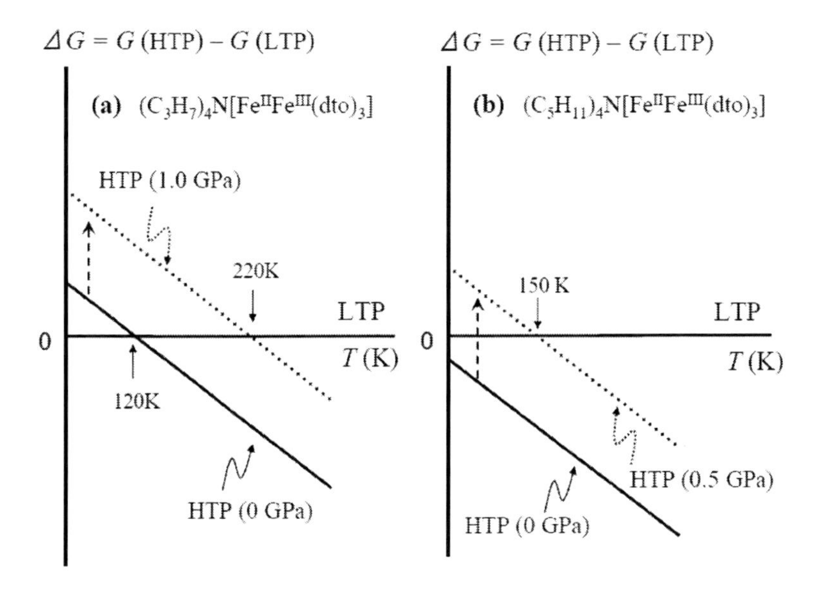

Figure 14. Schematic phase diagram of (a) $(C_3H_7)_4N[Fe^{II}Fe^{III}(dto)_3]$ and (b) $(C_5H_{11})_4N[Fe^{II}Fe^{III}(dto)_3]$. The vertical axis is the Gibbs energy relative to the LTP. The inclination of the HTP takes into account only spin entropy. The intercepts correspond to the difference of the enthalpy between the HTP and the LTP at 0 K.

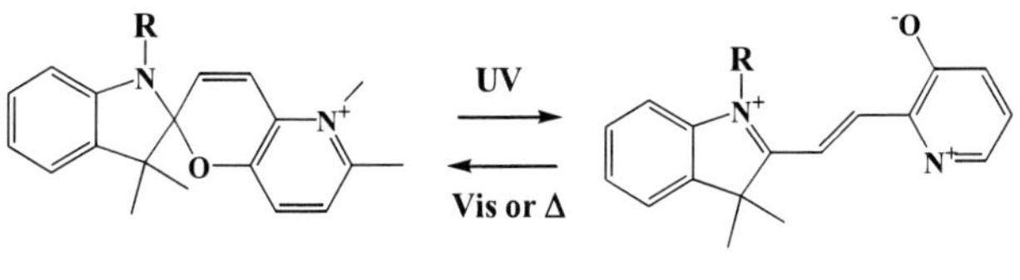

Figure 15. Photochromic reaction of spiropyran cation (SP-R).

Figure 16 show the change of the absorption spectra for (SP-Me)$[Fe^{II}Fe^{III}(dto)_3]$ in KBr pellet upon UV irradiation at 350 nm at 70 K. At 70 K, the absorption intensity around 570 nm is continuously enhanced with the increase of UV irradiation time while the initial black pellet is turned to deep purple one. After the UV irradiation (40 mW/cm^2) for 180 min, the intensity of the absorption spectrum corresponding to the π - π* transition of the OF is almost saturated. The UV light-induced OF in (SP-

Me)[$Fe^{II}Fe^{III}(dto)_3$] is stable even at room temperature in the dark condition. This absorption band almost disappears upon visible-light irradiation (600 mW/cm^2) for 120 min. The photo-isomerization of cationic SP-Me from CF to OF by UV irradiation and from OF to CF by visible-light irradiation reversibly takes place in (SP-Me)[$Fe^{II}Fe^{III}(dto)_3$] at 300 K as well as at 70 K.

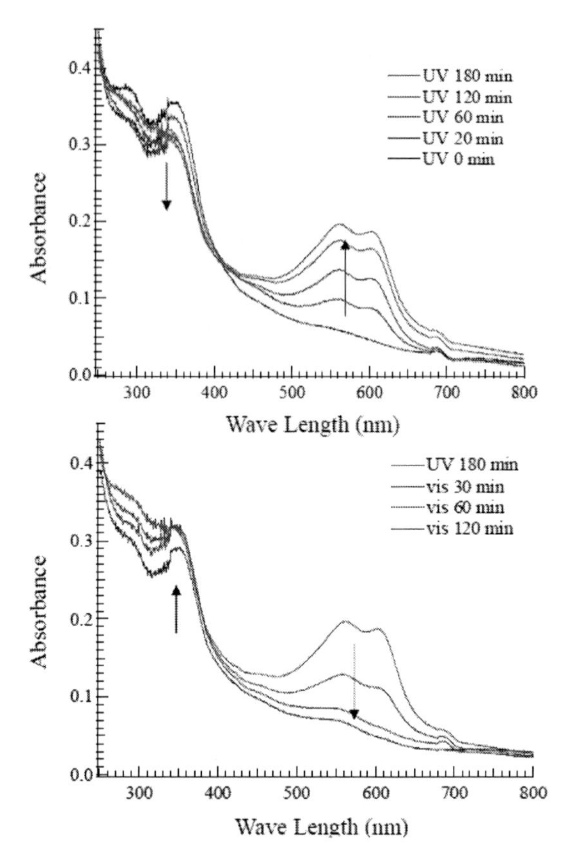

Figure 16. UV-vis absorption spectral change of (SP-Me)[$Fe^{II}Fe^{III}(dto)_3$] in KBr pellet upon UV irradiation and visible-light irradiation at 70 K. UV irradiation (350 nm, 40 mW/cm^2) on the pellet was carried out first. After the absorption spectra were saturated, visible-light irradiation (600 mW/cm^2) on the pellet was carried out.

(SP-Me)[$Fe^{II}Fe^{III}(dto)_3$] undergoes the CTPT as well as in the case of (n-$C_nH_{2n+1})_4N$[$Fe^{II}Fe^{III}(dto)_3$] (n = 3 and 4) in the similar temperature range. [29] When the temperature is decreased below 50 K, χT rapidly increases up to a maximum value around 18 K and the magnetization is saturated below that temperature, which suggests that (SP-Me)[$Fe^{II}Fe^{III}(dto)_3$] exhibits a long-range ferromagnetic ordering as well as $(n\text{-}C_nH_{2n+1})_4N[Fe^{II}Fe^{III}(dto)_3]$. In order to confirm and characterize the ferromagnetically ordered phase, the field cooled magnetization (FCM), the remnant magnetization (RM), and the zero-field cooled magnetization (ZFCM) were measured. These results are shown in Figure 17.[29] The FCM curve decreases stepwise at about 7 K and disappears at about 25 K. The RM curve also decreases stepwise at about 7 K and then disappears at about 22 K. The ZFCM curve, on the other hand, has two maxima at 5 K and 18 K. This peculiar behavior of magnetization curves is quite similar to that of $(n\text{-}C_4H_9)_4N[Fe^{II}Fe^{III}(dto)_3]$ in which the LTP and HTP coexist even in the temperature region below the CTPT. In analogy with $(n\text{-}C_4H_9)_4N[Fe^{II}Fe^{III}(dto)_3]$,[25] the LTP and HTP of (SP-Me)$[Fe^{II}Fe^{III}(dto)_3]$ individually undergo the ferromagnetic phase transitions with T_C (LTP) = 5 K and T_C (HTP) = 22 K, respectively.

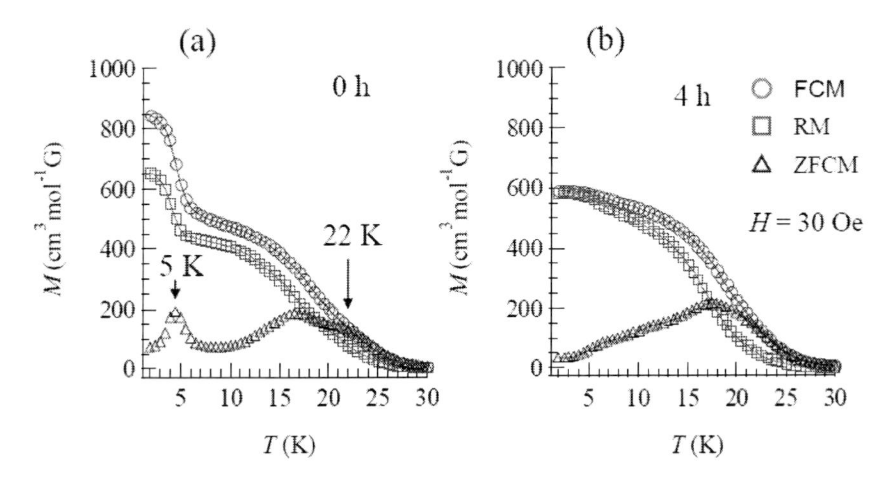

Figure 17. Temperature dependence of the magnetization for (SP-Me)$[Fe^{II}Fe^{III}(dto)_3]$ before and after the UV irradiation (4 hours, 40 mW/cm^2) at room temperature. FCM: field cooled magnetization, RM: remnant magnetization, ZFCM: zero-field cooled magnetization.

Moreover, the ferromagnetism of (SP-Me)$[Fe^{II}Fe^{III}(dto)_3]$ shows a noteworthy response upon UV irradiation. When the UV light of 350 nm is irradiated for 4 hours, the FCM and RM curves change from the stepwise magnetization curves to normal magnetization curves, and the peak around 5 K in ZFCM disappears, indicating the disappearance of the LTP. On the other hand, the FCM and RM values between 5 and 22 K are slightly increased after the UV irradiation. This photo-induced effect can be explained as follows. The photo-isomerization of cationic SP-Me from CF to OF by UV irradiation leads to the expansion of its own volume, which gives a significant stress to the framework of $[Fe^{II}Fe^{III}(dto)_3]$ layers and expands the unit cell volume. Taking into account that the CTPT in $(n\text{-}C_nH_{2n+1})_4N[Fe^{II}Fe^{III}(dto)_3]$ tends to be inhibited by the expansion of their $(n\text{-}C_nH_{2n+1})_4N^+$ cation size, the photo-isomerization of cationic SP-Me from CF to OF by UV irradiation presumably stabilizes the HTP and destabilizes the LTP, which induces the CTPT in the $[Fe^{II}Fe^{III}(dto)_3]$ layers. The photoisomerization-induced CTPT for (SP-Me)$[Fe^{II}Fe^{III}(dto)_3]$ is schematically represented in Figure 18. [29]

This new type of photomagnetism coupled with spin, charge and photon is triggered by a chemical pressure effect generated from the photo-isomerization of spiropyran from the CF to the OF. This phenomenon seems to have a significant similarity with the first events in the perception of light in rhodopsin where the photoisomerization of 11-cis-retinal into all-trans-retinal induces a conformational change in opsin and activates the associated G protein and triggers a second messenger cascade.

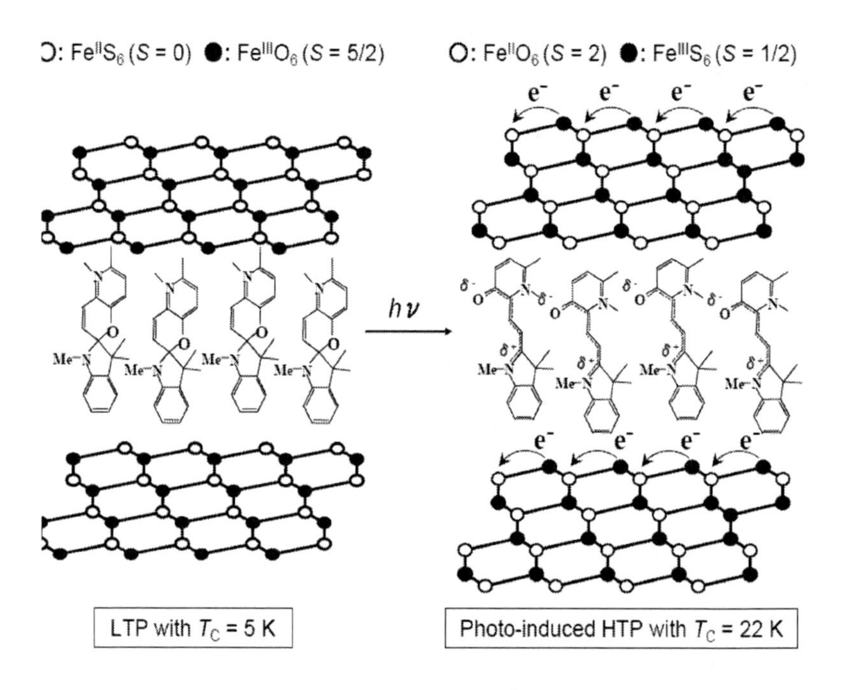

Figure 18. Schematic representation of the concerted phenomenon coupled with the CTPT in the $[Fe^{II}Fe^{III}(dto)_3]$ layers and the photo-isomerization of spiropyran in (SP-Me)$[Fe^{II}Fe^{III}(dto)_3]$. White colored and black colored circles are the Fe^{II} and Fe^{III} sites, respectively.

CONCLUSION

Magnetic coordination polymers can provide various controllable magnetic properties due to the intrinsic tunability of both the electronic and structural states. In this chapter, the development of photomagnetic organic-inorganic hybrid assemblies was summarized. The photomagnetism was successfully designed by intercalation of the photochromic organic molecules into magnetic inorganic arrays. However, further studies are required to achieve more significant photo-switch/control of magnetism. For example, photomagnetic behavior at higher temperature (*ca.* room temperature) is indispensable for practical uses. The conversion ratio of the photochromic molecule in the magnetic inorganic arrays should be enhanced in order to achieve more drastic control/switch. The versatility of organic-inorganic hybrid assemblies would effectively contribute toward tackling these problems.

ACKNOWLEDGMENTS

It is a great pleasure to acknowledge the followings for their significant contributions to this work: M. Enotomo, A. Okazawa, H. Shimizu, Y. Ono, M. Itoi, N. Kida, M. Hikita, I. Kashima, W. Aoki,

REFERENCES

[1] R. Boca, *Theoretical foundations of molecular magnetism*, Elsevier Science S.A., (1999).
[2] L. Cambi, A. Cagnasso, *Atti Acad. Naz. Lincei*, (1931) 13, 809.
[3] O. Kahn, *Molecular Magnetism*, Wiley-VCH Inc., (1993), p. 160.
[4] S. Decurtins, P. Gütlich, C. P. Köhler, H. Spiering, A. Hauser, *Chem. Phys. Lett.*, (1984) 105, 1.
[5] O. Sato, S. Hayami, Z. Z. Gu, K. Takahashi, R. Nakaima, K. Seki, A. Fujishima, *J. Photochem. Photobiol.* A, (2002) 149, 111.
[6] M. Nihei, Y. Sekine, N. Suganami, K. Nakazawa, A. Nakao, H. Nakao, Y. Murakami, H. Oshio, *J. Am. Chem. Soc.*, (2011) 133, 3592.
[7] S. Ohkoshi, K. Imoto, Y. Tsunobuchi, S. Takano, H. Tokoro, *Nat. Chem.*, (2011) 3, 564.

[8] M. Ogawa, K. Kuroda, *Chem. Rev.*, (1995) 95, 399.
[9] P. Day, A. E. Underhill, Ed., *Metal-Organic and Organic molecular magnets.,* The Royal Society, (1999).
[10] V. Laget, C. Hornick, P. Rabu, M. Drillon, R. Ziessel, *Coord. Chem.* Rev., (1998) 178-180, 1533.
[11] W. Fujita, K. Awaga, *Inorg. Chem.*, (1996) 35, 1915.
[12] W. Fujita, K. Awaga, T. Yokoyama, *Inorg. Chem.*, (1997) 36, 196.
[13] M. Okubo, M. Enomoto, N. Kojima, *Solid State Commun.*, (2005) 134, 461.
[14] H. Shimizu, M. Okubo, A. Nakamoto, M. Enomoto, N. Kojima, *Inorg. Chem.*, (2006) 45, 10240.
[15] J. C. Crano, R. J. Guglielmetti, Ed. *Organic photochromic and thermochromic compounds, Plenum Press*, (1999).
[16] K. Matsuda, M. Irie, *J. Am. Chem. Soc.*, (2000) 122, 7195.
[17] N. Kojima, M. Okubo, H. Shimizu, M. Enomoto, *Coord. Chem. Rev.,* (2007) 251, 2665.
[18] N. Kojima, W. Aoki, M. Itoi, M. Seto, Y. Kobayashi, Yu. Macda, *Solid State Commun.*, (2001) 120, 165.
[19] T. Nakamoto, Y. Miyazaki, M. Itoi, Y. Ono, N. Kojima, M. Sorai, Angew. *Chem. Int. Ed.*, (2001) 40, 4716.
[20] M. Itoi, A. Taira, M. Enomoto, N. Matsushita, N. Kojima, Y. Kobayashi, K. Asai, K. Koyama, T. Nakamoto, Y. Uwatoko, J. Yamamura, *Solid State Commun.*, (2004) 130, 415.
[21] N. Kojima, Y. Ono, Y. Kobayashi, M. Seto, *Hyperfine Interact.*, (2004) 156-157, 175.
[22] T. Birchall, K. M. Tun, *Inorg. Chem.*, (1976) 15, 376.
[23] S. G. Carling, D. Visser, D. Hautot, I. D. Watts, P. Day, J. Ensling, P. Gütlich, G. L. Long, F. Grandjean, *Phys. Rev. B*, (2002) 66, 104407.
[24] R. I. Collins, R. Pettit, W. A. Baker Jr., *J. Inorg. Nuclear Chem.*, (1966) 28, 1001.
[25] M. Itoi, Y. Ono, N. Kojima, K. Kato, K. Osaka, M. Takata, Eur. *J. Inorg. Chem.*, (2006), 1198.
[26] T. Nakamoto, Y. Miyazaki, M. Itoi, Y. Ono, N. Kojima, M. Sorai, *Angew. Chem. Int. Ed.*, (2001) 40, 4716.
[27] M. Sorai, S. Seki, *J. Phys, Chem. Solids*, (1974) 32, 555.
[28] Y. Kobayashi, M. Itoi, N. Kojima, K. Asai, *J. Phys. Soc. Jpn.*, (2002) 71, 3016.
[29] N. Kida, M. Hikita, I. Kashima, M. Okubo, M. Itoi, M. Enomoto, K. Kato, M. Takata, N. Kojima, *J. Am. Chem. Soc.*, (2009) 131, 212.

In: Research Advances in Magnetic Materials ISBN: 978-1-62417-913-6
Editors: C. Toulson and D. Marwick

Chapter 4

SPIN-PREFERENCE RULES FOR CONJUGATED POLYRADICALS

Masashi Hatanaka*
Department of Green and Sustainable Chemistry, School of Engineering,
Tokyo Denki University, Adachi-ku, Tokyo, Japan

ABSTRACT

Since 1950's, prediction of spin-quantum number in conjugated radicals has attracted many theoretical chemists. There have been two major trends in spin-preference rules of conjugated biradicals. One is based on molecular orbital (MO) theory supported by Hund's rule. Another is based on valence bond (VB) theory, which has been intuitive for many chemists as a spin-polarization rule with spin alternation. Nowadays, as revealed experimentally and theoretically, localized orbital methods are better than the classical MO or VB methods in that some exceptional radicals that violate the simple MO or VB predictions are correctly described. Modern formalism of the spin-preference rule deals with exchange integrals between maximally localized MOs, which have been expanded from biradicals to general polyradicals by using Wannier functions. It has been shown that the best orbitals for description of the spin states of polyradicals are maximally localized Wannier functions, which minimize the exchange integrals. The Wannier analysis has been applied to many polyradicals and supported by semi-empirical, *ab initio*,

* E-mail:mhatanaka@xug.biglobe.ne.jp.

and DFT calculations. In this chapter, construction and use of the Wannier-function method are reviewed.

1. INTRODUCTION

Spin preference of conjugated biradicals has attracted many theoretical chemists. Longuet-Higgins formulated a spin preference rule of conjugated biradicals by the Hückel molecular orbital (MO) method and Hund's rule [1]. Conjugated biradicals often have degenerate frontier orbitals. The degenerate orbitals are non-bonding type, of which eigenvalues are zero with respect to the Coulomb integral α. These are called non-bonding molecular orbitals (NBMOs). In most cases (apart from some exceptions as shown later), the number of the NBMOs can be written as (N_c - $2T_d$), where N_c and T_d are numbers of the carbon atomic sites and classical double bonds, respectively [1]. From the Hund's rule, the spin-quantum number at the ground state becomes (N_c - $2T_d$)/2. Thus, classical theories on the spin preference have been related to the number of NBMOs, in which all the bonding electrons are spin-paired below the frontier levels. One can predict the spins at the ground state by counting the number of NBMOs and applying the Hund's rule. Figure 1 shows famous biradicals trimethylenemethane (*1*), tetramethyleneethane (*2*), cyclobutadiene (*3*), *m*-phenylene (*4*), pentamethylenepropane (*5*), and tetramethylenebenzene (*6*). The dots represent radical centers created by methylene or methine groups. According to the prediction above, all the biradicals except *3* have two usual NBMOs. *3* has no radical centers. However, actually, *3* has two unusual NBMOs. NBMOs of *3* arise from accidental degeneracy of 4n annulenes. Thus, from Hund's rule, all the biradicals above are predicted to be ground-state triplet. Actually, this is not correct. While *1* [2,3] and *4* [4] have triplet ground states, as established theoretically and experimentally, *2* [5], *3* [6], *5* [7,8], and *6* [9] have been proved to be singlet.

Another trend in the spin-preference rules is based on valence bond (VB) theory [10]. The carbon atomic sites in conjugated hydrocarbons are divided into so-called starred and unstarred atoms that are not adjacent each other. This theory supposes spin alternation between the starred and unstarred atoms corresponding to the maximally spin-paired resonance structures, and the residual spins become parallel due to the ferrimagnetic description. This is often referred as spin-polarization rule, which is essentially deduced from the Heisenberg model. The spin alignment results from spin polarization (spin alternation) through the each double bond, and thus, description of this rule should be interpreted along the VB context. If the numbers of starred atoms (N_s) and unstarred atoms (N_u) are same (N_s=N_u), the spin quantum number at the ground state is predicted to be zero. On the other hand, if (N_s-N_u) is not zero (N_s>N_u), the spin-quantum number at the ground state becomes $|N_s$-$N_u|/2$ [10]. We note that the physical meaning of this situation is different from that of the MO-based Hund's rule. Figure 2 shows spin polarizations for *1-6*. By counting numbers of up and down spins, we distinguish possible high- and low-spin ground states. Due to the ferrimagnetic images, *1*, *4*, *5* are predicted to be triplet. On the other hand, *2*, *3*, *6* are predicted to be singlet due to cancelation of the spins. We see that MO and VB predictions do not always coincide. Actual determinations of spin-quantum number at the ground states are done by experiments such as ESR or magnetic susceptibility measurements, or theoretical calculations by *ab initio* methods.

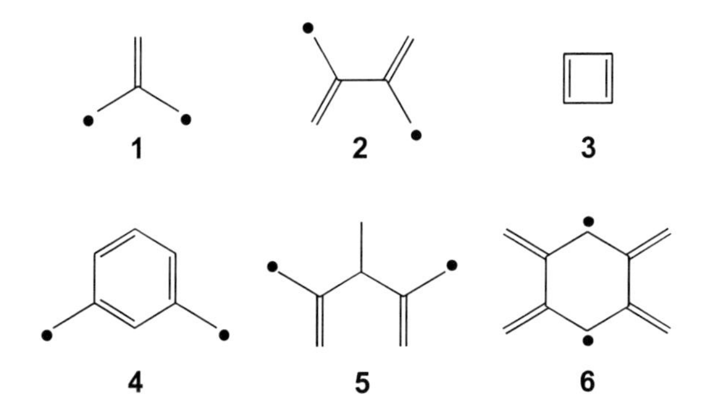

Figure 1. Molecular structures of typical biradicals *1-6*.

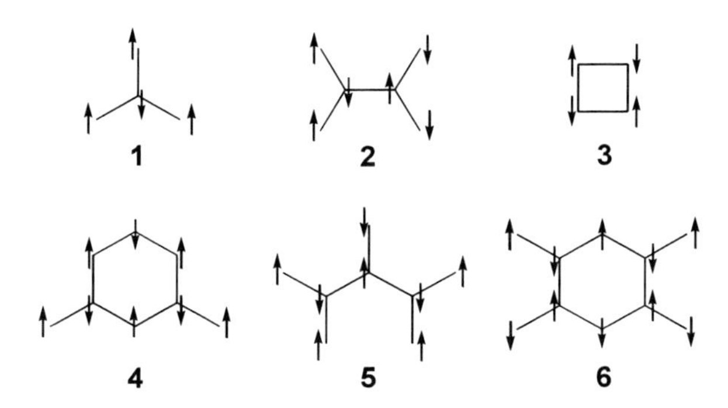

Figure 2. Scheme of spin polarization (spin alternation) for *1-6*.

Modern spin preference rules for conjugated radicals have been developed in studies on violation of the Hund's rule and spin-polarization rule. Tetramethyleneethane *2* has two NBMOs and two 'unpaired' electrons. Similarly, Cyclobutadiene *3* also has two NBMOs and two 'unpaired' electrons. From Hund's rule, these molecules are expected to be triplet. However, the ground states are calculated to be singlet at any modern *ab initio* level of theory. *2* with planar geometries prefers singlet [5], and *3* is also singlet molecule [6], regardless of the two NBMOs. That is, *2* and *3* are important exceptions for the Hund's rule. In addition, *3* is subject to pseudo Jahn-Teller effect to be distorted from D_{4h} to D_{2h} symmetry. Such vibronic interactions often stabilize low-spin states. However, these are additional problems with higher-order perturbations. Similarly, pentamethylenepropane *5* also has two NBMOs and two unpaired electrons. However, contrary to the Hund's rule, the ground state is singlet, which has been established experimentally and theoretically by using nitroxide derivatives [7,8]. Moreover, this molecule also violates the VB theory in that the spin-polarized (spin-alternated) description leads to ferrimagnetic ground state, that is, triplet. In this sense, pentamethylenepropane *5* is an important exception for both the Hund's rule and spin-polarization rule.

Borden and Davidson resolved these controversial situations [11]. They paid attention to the amplitude patterns of NBMOs in conjugated biradicals, and classified into two types. Though NBMOs are generally not determined uniquely due to the degeneracy, they found an important criterion to distinguish the ground-state spin states. If the two NBMOs remain to span common atoms after any linear combination, they are called non-disjoint, and the system is triplet. On the other hand, if the two NBMOs can be made to span no common atoms, they are called disjoint, and the system is singlet. Qualitative aspects and confirmations of this rule have been well reviewed by Borden et al [12,13]. This rule is based on evaluating the exchange integral between the NBMOs. In non-disjoint systems, the common-spanned atoms cause significant exchange integral due to the Pauli principle. Thus, electrons with parallel spins are stabilized by reduction of Coulomb repulsion. The Borden-Davidson rule has been mathematically established by using unitary transformation, of which parameter is the degree of freedom within the degenerate space [14]. Figure 3 shows schematic representations of the best unitary-transformed NBMOs. In general, the best orbitals are not canonical types but localized types. From the Borden-Davidson definitions, NBMOs of *1* and *4* are non-disjoint, and NBMOs of *2*, *3*, *5*, and *6* are disjoint. The biradicals in the former group have triplet ground states, which have been well established. The biradicals in the latter group have singlet ground states. Modern precise calculations and experiments including their stable derivatives have proved their singlet ground states [12,13].

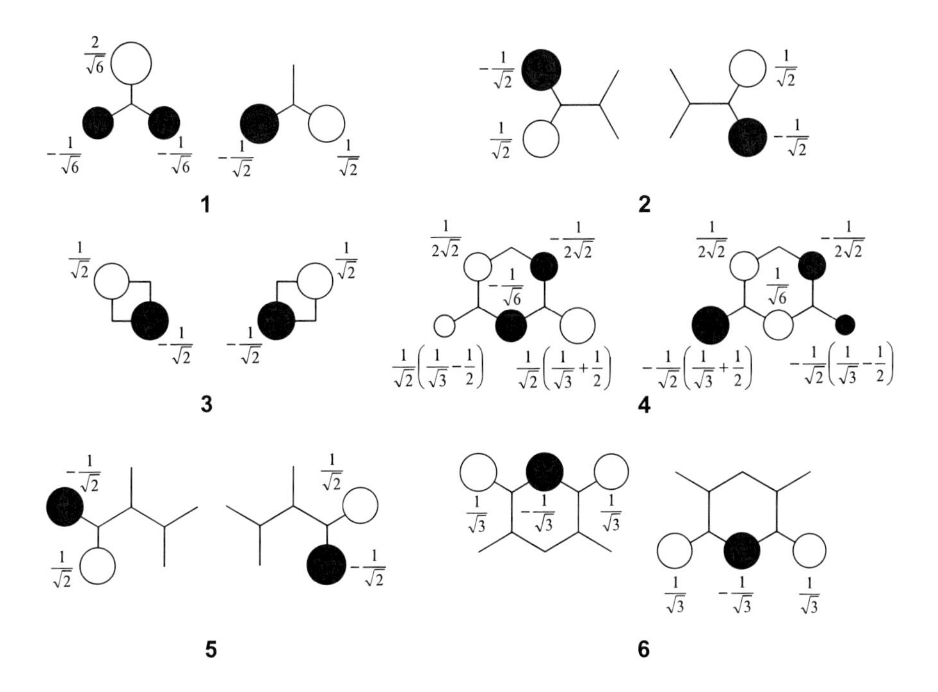

Figure 3. Amplitude patterns of NBMOs for *1-6*.

In recent, the localized orbital method based on the Borden-Davidson rule has been applied to not only planar molecules but also face-to-face stacked radicals [15]. Figure 4 shows some face-to-face stacked radicals. *7* is an allyl-radical dimer, in which 1- and 2-positions are faced through pantogragh-like linkage. *8* and *9* are benzyl-radical dimers, in which two benzyl radicals are faced in pseudo *para*- and *ortho*- fashions, respectively. In these geometries, *7-9* have pseudo non-disjoint NBMOs, and their ground states are predicted to be triplet [15]. Experimentally, cyclophane-linked carbene derivatives of *8* and *9* have been proved to have high-spin ground states (quintet), and the pseudo-*meta* derivatives have been proved to be antiferromagnetic (singlet) [16,17]. Classically, spin preference of these π-stacked radical assemblies has been predicted by McConnell's rule based on Heisenberg model [18], similar to the VB theory by Ovchinnikov [10]. For *7-9*, one can describe ferrimagnetic schemes with spin alternation at each adjacent site by VB description.

However, nowadays, the localized orbital method is probably better than the classical views in that some radical assemblies violate such VB descriptions, as shown later. In this sense, π-stacked radicals can be regarded as extended non-Kekulé molecules in that their skeletons are mathematically reduced to the topological linkage, similar to planar systems. In π-stacked radicals, intra- and inter-molecular resonance integrals are generally different. Even if signs of intra-molecular resonance integrals β are always minus, signs of inter-molecular resonance integrals β' are not a priori determined. However, expressions of the exchange integral do not depend on signs of the resonance integral. Supposing that β and β' are constant, exchange integrals are functions of ratio $|\beta'/\beta|$, which depends only on the intermolecular distance between the faced moieties. Thus, under proper topological conditions, we can deal with the π-stacked radicals as non-bonding degenerate systems with non-Kekulé character. Interestingly, for a given π-stacked dimer, there exists an optimum intermolecular distance that maximize the high-spin stability. This theorem can be proved by using localized orbitals as functions of the intermolecular distance [15]. This cannot be deduced from any classical theory. The proof is based on unitary transformation of the localized orbitals, and also independent of signs of resonance integrals. The optimized intermolecular distance is ca. 2.0 Å [15]. If the optimum points on high-spin stabilities are observed by experiments, validity of the localized orbital methods will be further established.

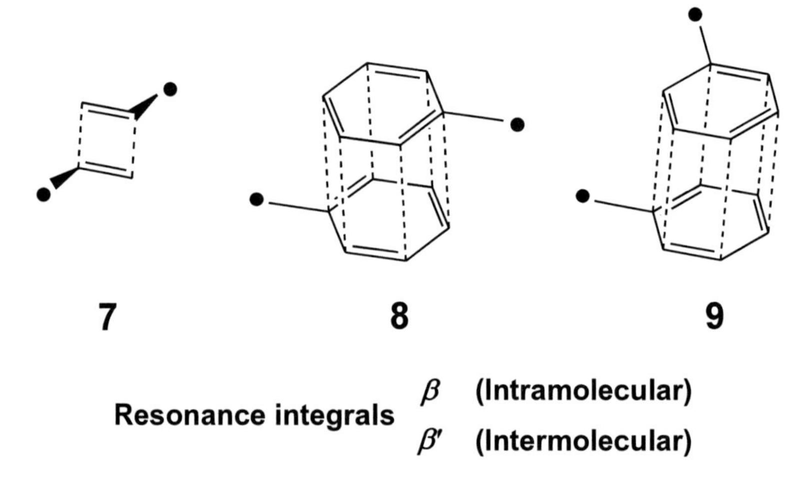

Figure 4. π-stacked allyl and benzyl dimers *7-9*.

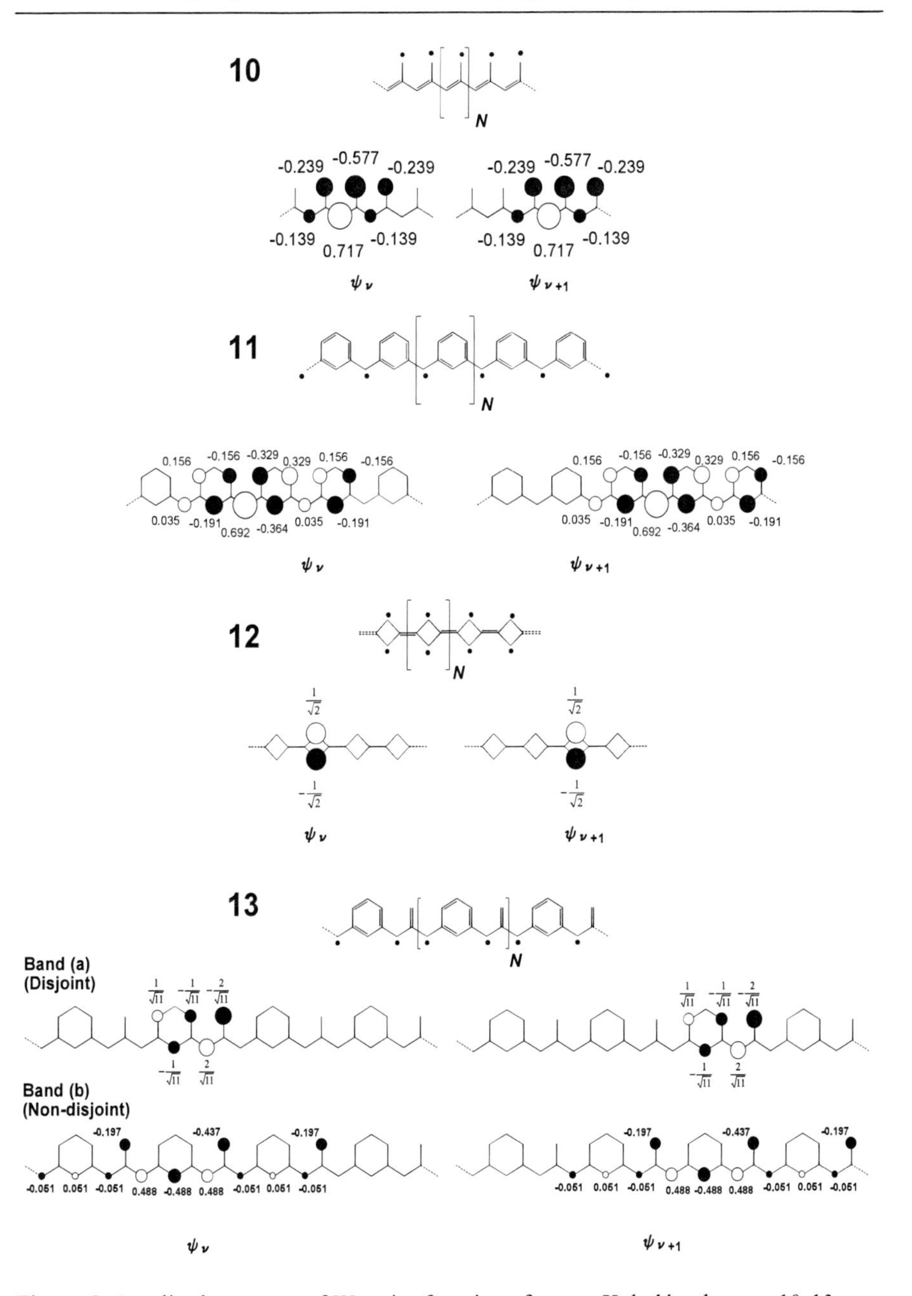

Figure 5. Amplitude patterns of Wannier functions for non-Kekulé polymers *10-13* (excerpted from references [19] (for *10-12*) and [21] (for *13*)).

2. SPIN PREFERENCE OF POLYRADICALS

As for polyradicals, formulation of the spin-preference rule has been long prevented due to the infinite-fold degeneracy of the NBMOs. In non-Kekulé polyradicals, NBMOs form bands with zero width as an assembly of non-bonding crystal orbitals (NBCOs). In principle, direct calculation of exchange integrals between each NBCO is possible, but the resultant electronic states (particularly in low-spin states) are not always optimized at the variational space. Due to degree of freedom within the degeneracy, we need to minimize each exchange integral by a proper unitary transformation. Evaluation of exchange integrals between each NBCO are done by Wannier transformation. The formulation is not so complicated, as follows [19].

Let us consider Bloch functions corresponding to a given NBCO band:

$$\varphi_k = \frac{1}{\sqrt{N}} \sum_{\mu}^{N} \sum_{r}^{cell} \exp(ik\mu) C_r(k) \chi_{\mu,r}, \tag{1}$$

where the wavenumber k runs from $-\pi$ to π, μ is the cell index, N is the number of cells, and r is the index of atomic orbitals (usually, p_z orbitals of carbon). In general, $C_r(k)$ is complex number. However, we adopt the real part of $C_r(k)$ in order to minimize the exchange integral of the system:

$$C'_r(k) = \frac{1}{2}\left\{C_r(k) + C_r(k)^*\right\} = \frac{1}{2}\left\{C_r(k) + C_r(-k)\right\}. \tag{2}$$

The Wannier function localized at the ν-th cell is defined by Equations (3) and (4):

$$\psi_\nu = \sum_{\mu}^{N} \sum_{r}^{cell} a_r(\mu - \nu) \chi_{\mu,r}, \tag{3}$$

$$a_r(\mu - \nu) = \frac{1}{2\pi} \int_{-\pi}^{\pi} \exp\{i(\mu - \nu)k\} C'_r(k) dk, \tag{4}$$

where

$$\tau = \mu - \nu. \tag{5}$$

This is a special unitary transformation for infinite-fold degenerate orbitals. Each Wannier coefficient is a function of the integer τ, which represents the difference from the ν-th cell. We note that Wannier functions deduced from real part of $C_r(k)$ are not always normalized into unity. The Wannier functions should be normalized using a proper normalization factor. Under the Hückel approximation, we can adopt the normalization factor:

$$C' = \frac{1}{\sqrt{\sum_{\tau}\sum_{r}\left|a_r(\tau)\right|^2}}. \tag{6}$$

In general, Wannier functions decay very rapidly with respect to distance from the ν-th Wannier center. That is, each Wannier function coefficient $a_r(\tau)$ localizes only at one or a few unit cells around the ν-th cell. Therefore, $a_r(\tau)$ with $|\tau|\geq 2$ can be ignored, and the exchange integral K_{ij} between the i-th and j-th Wannier functions is nontrivial only when $|i\text{-}j| = 1$:

$$\begin{aligned}
K_{ij} &= \iint \psi_i(1)\psi_j(1)\frac{e^2}{r_{12}}\psi_i(2)\psi_j(2)\,d\tau_1 d\tau_2 \\
&= \sum_r\sum_s\sum_t\sum_u a_r(i)a_s(j)a_t(i)a_u(j)(rs|tu) \\
&\cong \left\{\sum_r^{i-th\,cell}\left\{a_r(0)^2 a_r(-1)^2\right\} + \sum_r^{j-th\,cell}\left\{a_r(0)^2 a_r(1)^2\right\}\right\}(rr|rr) \\
&\propto 2\sum_r^{i-th\,cell}\left\{a_r(0)^2 a_r(1)^2\right\},
\end{aligned} \tag{7}$$

where $(rs|tu)$ denotes the electron-repulsion integrals, and only one-centered integrals were taken into account in the approximation. The factor 2 in the last expression results from Equation (2), which guarantees even-function properties of the resultant Wannier functions. Thus, we can deduce the exchange integrals in the given NBCO from amplitude pattern of the Wannier functions. From the variation principle using complex phase factors, we can prove that the Wannier functions with even-function properties above minimize the total exchange integral of the system, and the set of Wannier funtions is unique [19]. Then, they are maximally localized, because the exchange integral is related to degree of localization. That is, disjoint property of Wannier functions is approximately proportional to degree of localization. The degree of localization resembles that of maximally localized Wannier functions [20]. If the best Wannier functions span common atoms, the exchange integrals are positive. If the best Wannier functions span no common atoms, the exchange integrals are zero.

Now we can predict the spin states of general polyradicals by whether or not the adjacent Wannier functions span common atoms. If the best Wannier functions span common atoms, the system becomes ferromagnetic with significant positive exchange interactions. This is akin to non-disjoint biradicals. On the other hand, if the best Wannier functions span no common atoms, the system becomes antiferromagnetic with zero exchange interactions. This is akin to disjoint biradicals. Usually, in the latter case, through-space interactions between the disjoint (that is, spatially separated) Wannier functions lead to antiferromagnetic ground states. The best Wannier function is maximally localized type that is intuitive to chemists. In this sense, the Borden-Davidson rule and Wannier function method can be called as localized orbital method.

3. Wannier Analysis of Polyradicals

Non-bonidng Wannier functions of typical non-Kekulé polyradicals are shown in Figure 5. *10*, *11*, and *12* are prototypes for conjugated polyradicals that have been well studied by oligomer models or band theory. We see that while in *10* and *11* the Wannier functions span common atoms between the adjacent cells, in *12*, the Wannier functions span no common atoms. Therefore, *10* and *11* have non-disjoint NBCOs and predicted to be ferromagnetic. *12* has disjoint NBCOs and predicted to be antiferromagnetic. The magnitudes of exchange integrals can be estimated by quantum-chemical calculations, as shown later. Interestingly, disjoint/non-disjoint composite systems are also constructed. The typical example is *13* [21]. This has both disjoint and non-disjoint bands, as seen from the two independent Wannier functions. It is not difficult to show the total exchange integrals are non trivial and the system becomes essentially non-disjoint. Indeed, *13* is predicted to be ferromagnetic by DFT calculations, as shown later. That is, in disjoint/non-disjoint composite systems with plural radical centers, at least one non-disjoint band leads to positive exchange integrals.

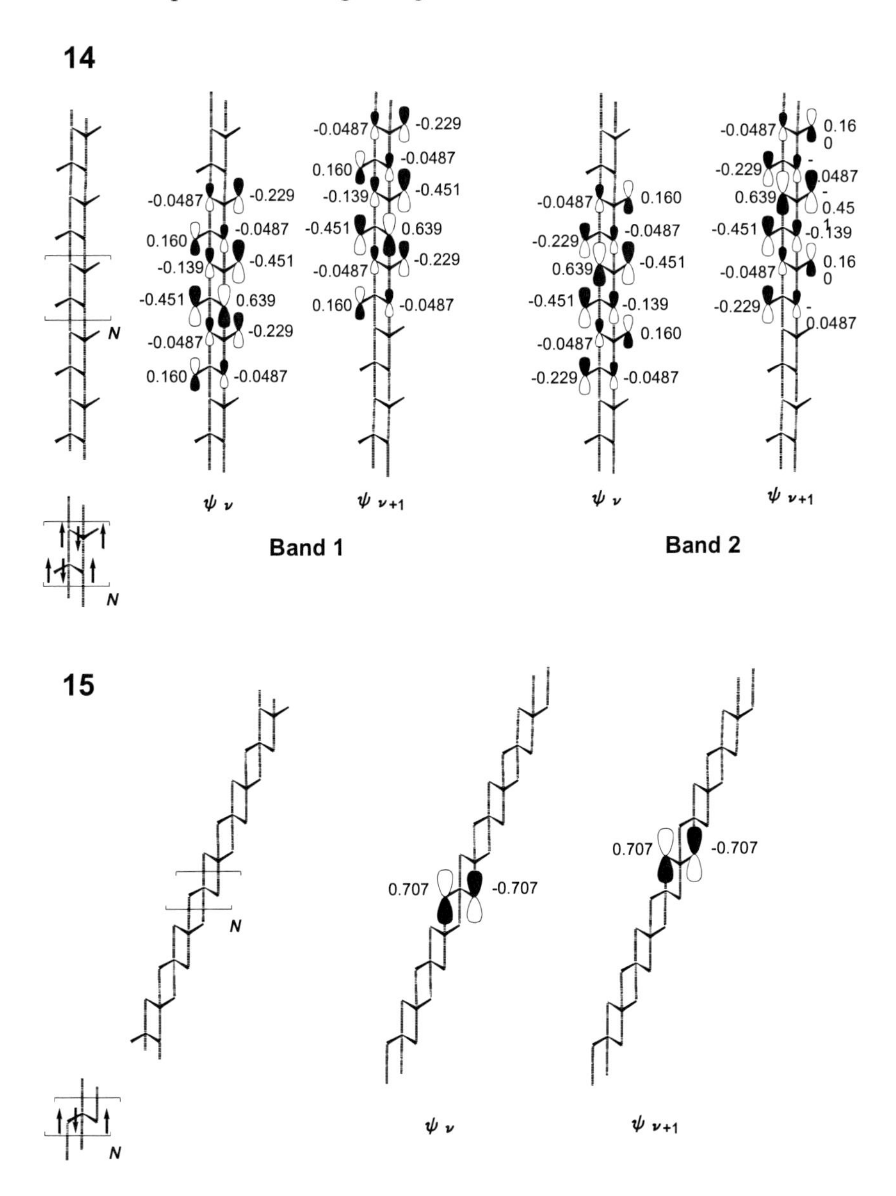

Figure 6. Amplitude patterns of Wannier functions for allyl radical assemblies *14* and *15* (excerpted from reference [19]).

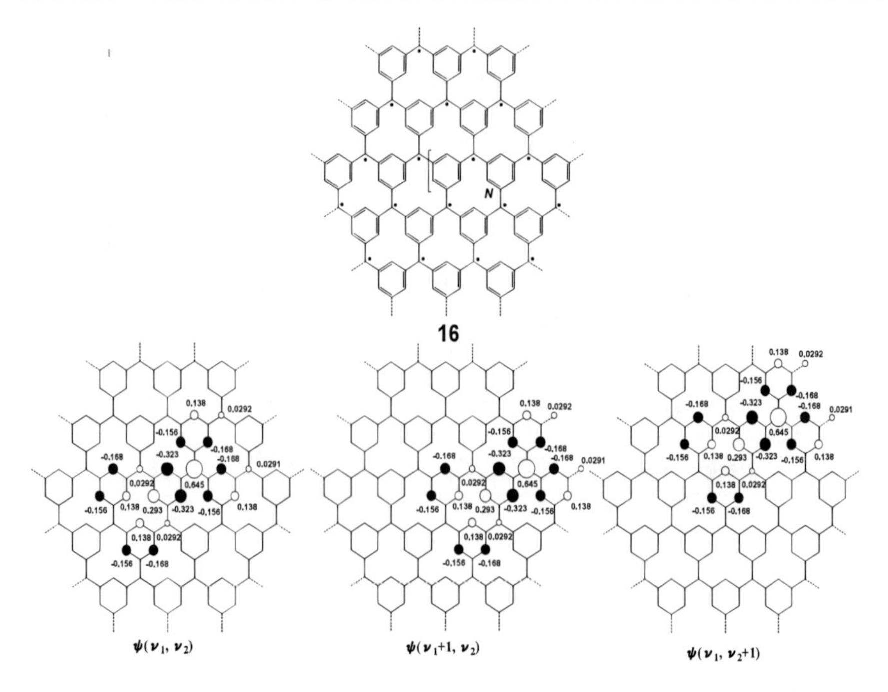

Figure 7. Amplitude patterns of Wannier functions for Mataga polymer *16* (excerpted from reference [19]).

In biradicals, some exceptional species that violate the early methods leaded to the correct spin-preference rule. In polyradicals, how can we show the reasonable proof for the Wannier function method? Validity of the Wannier function method is also shown by using special cases that violate classical rules. Here we consider two types of allyl radical assemblies, as shown in Figure 6. One is the ladder-like assembly *14*, and another is magic-hand-like assembly *15* [22]. If only topological linkages are considered, *14* and *15* have non-bonding NBCOs, similar to usual non-Kekulé polymers. So, from the Hund's rule, high-spin ground states are predicted in both cases. In addition, from the spin-polarization rule (see the arrow schemes in Figure 6), both *14* and *15* are predicted to be ferromagnetic. However, as shown in Figure 6, their Wannier functions do not always cause positive exchange integrals. That is, while NBCOs of *14* are non-disjoint, NBCOs of *15* are disjoint. We note that *14* has two independent bands at the frontier levels due to the plural radical centers. Thus, from the Wannier function method, *14* is predicted to be ferromagnetic, and *15* is predicted to be antiferromagnetic. In this sense, *15* is an important exception for classical MO theory and VB theory. Exchange integral per unit is obtained by using oligomer models. That is, total spin gaps (energy gaps between the lowest- and highest-spin states) divided by the oligomer length can be regarded as magnitude of the ferromagnetic or antiferromagnetic interactions. In one-dimensional systems, effective exchange integrals are obtained from the spin gaps divided by the number of radical pairs. The effective exchange integrals are not always positive. Positive and negative exchange integrals represent ferromagnetic and antiferromagnetic interactions, respectively. Comparative summary of the effective exchange integrals are shown later.

4. *Ab initio* Studies on Polyradicals

Recent *ab initio* studies on spin preference of polyradicals are fully consistent with the formulation above. Degenerate systems particularly in low-spin states cannot be described only by one Slater determinant. Hartree-Fock wavefunctions with one Slater determinant are often inadequate even for qualitative prediction of spin preference. Indeed, if multi configurations are included, semi-empirical wavefunctions are often better than *ab initio* Hartree-Fock wavefunctions with one Slater determinant. Nowadays, CI (configuration interactions) or DFT (density-functional theory) is convenient in that electronic correlations are easily included for description of the degenerate systems. Perturbation theory is also effective for electronic correlations, though the actual calculations often take very long time. In these methods, multi configurations required are automatically included by the electronic correlations. Spin densities are calculated by UHF (unrestricted Hartree-Fock) wavefunctions. Under proper periodic boundary conditions, qualitative analysis on spin densities can be done for the typical polyradicals [23]. The critical temperature is also given in principle [24]. However, the calculated spin densities do not always correspond to the classical VB descriptions with spin alternation. In addition, we note that UHF wavefunctions do not satisfy spin symmetry precisely. Therefore, unrestricted wavefunctions are acceptable only when the expectation values of spin square are close to those of the pure multiplet states. Apart from the spin symmetry, recent advances in DFT have also been applied to band calculations under periodic boundary conditions [25], as well as oligomer models. Strictly speaking, DFT formulation should be expanded to multi-reference calculations. However, at present, this is too complex and expensive for our purpose. Though precise calculations are important problems, theoretical importance lies in qualitative description of spin preference, rather than quantitative details.

Table 1. Comparative summary of spin preference rules

Classical MO theory[a)] (Hund's rule)	VB (spin-polarization) theory[a)] (Heisenberg spin alternation)	Localized orbital method[a)] (Wannier function)	Effective exchange integrals[b)] (kcal/mol per radical center)
1. FM	FM	FM	(+16.1±0.1 [c)])/2
2. FM	AFM	AFM	-2.8 [d)]/2
3. FM	AFM	AFM	-8.4 [e)]/2
4. FM	FM	FM	+10.0 [f)]/2
5. FM	FM	AFM	-5.1×10^{-3} [g)]/2
6. FM	AFM	AFM	-4.6 [h)]/2
7. FM	FM	FM	+27.5 [i)]/2
8. FM	FM	FM	+22.3 [j)]/2
9. FM	FM	FM	+21.1 [k)]/2
10. FM	FM	FM	+24.5 [l)]
11. FM	FM	FM	+4.6 [m)]
12. FM	AFM	AFM	-9.3 [n)]
13. FM	FM	FM	+11.3 [o)]
14. FM	FM	FM	+3.8 [p)]
15. FM	FM	AFM	-3.9 [q)]

[a)] Qualitative predictions as ferro (or ferri) magnetic (FM) and antiferromagnetic (AFM) ground states.

[b)] Effective exchange integrals in biradicals are expressed by the singlet-triplet energy gaps divided by 2.

[c)] Experimental value by PES spectra [3].

[d)] SDCI/ Dunning spilt-valence basis set [5].

[e)] MR-SDCI-(Q)/6-31G*+ΔZPE at D_{4h} symmetry [6].

[f)] SDTQ-CI/Dunning spilt-valence basis set [4].

[g)] Experimental value for a nitroxide derivative by ESR and SQUID [7,8].

[h)] CASPT2N/6-31G*//UHF/6-31G* [9].

[i)] PM3-CI(4,4) with intermolecular distance d=2.0Å [15].

[j)] PM3-CI(4,4) with intermolecular distance d=2.0Å [15].

[k)] PM3-CI(4,4) with intermolecular distance d=1.8Å [15].

[l)] VWN/6-31G* [19].

[m)] BLYP-CO/4-31G [25].

[n)] PM3-CI(8,8) [21].

[o)] VWN/6-31G* [21].

[p)] B3LYP/3-21G [22]. q) B3LYP/3-21G [22].

Table 1 is shows calculated exchange integrals for biradicals considered. Effective exchange integrals in biradicals are expressed by the singlet-triplet energy gaps divided by 2. Comparative summary for the classical MO, VB and localized orbital methods is also shown. FM and AFM denote ferromagnetic (ferrimagnetic) and antiferromagnetic interactions predicted by the spin-preference rules. Actual sign of the magnetic interactions are determined by effective exchange integrals calculated from proper quantum-chemical calculations. Effective exchange integrals with plus and minus signs correspond to ferromagnetic and antiferromagnetic interactions, respectively. Nowadays, most of the effective exchange integrals are available at post-Hartree-Fock and/or DFT level of theories. We see that there are some exceptional situations in the classical MO theory. In VB theory, we also see an important exception *5* that violates the qualitative rule. It can be seen that the localized orbital method gives fully correct predictions, consistent with the sign of the exchange integrals.

Table 1 also shows calculated exchange integrals for polyradicals and allyl-radical assemblies above. Validity of the localized orbital method (Wannier function method) is seen from the qualitative and quantitative spin preference. We see that an important exception *15* that violates the classical theory obeys the localized orbital (Wannier function) method. Thus, we can conclude that localized orbital methods are better than the classical MO and VB descriptions. The Wannier functions described above are not only expansion of Borden-Davidson localized orbitals but also bridges to physical descriptions of organic ferromagnets. From these calculations, we can imagine that the Wannier function method is available for not only complete degenerate systems but also nearly degenerate systems. This is important for actual designing for the organic ferromagnets. In principle, Wannier functions can be constructed for any band with any band width. When the band width is enough small, the frontier bands should be half occupied, and resultant high-spin states should be more stable than the closed-shell low-spin states. This situation becomes important in hetero-atom containing systems or spin-doped systems, because hetero atoms or dopants perturb the original degeneracy. On the other hand, when the band width is larger than the exchange integrals, low-spin states with closed-shell should be more stable than the high-spin states. It is noteworthy that flatness of frontier bands itself is not the sufficient condition but necessary condition for the ferromagnetic interactions.

5. Expansion to Two- and Three-dimensional Systems

The Wannier function method is easily expanded to two- and three-dimensional systems. In view of actual synthesis, magnetism of one-dimensional systems is sensitive to defects, because there is only one path to control the ferromagnetic interactions. So, in order to obtain practically usable magnets, we need to design two- or three-dimensional systems. Two-dimensional systems are very important in relation to experimental data. For three-dimensional systems, intermolecular ferromagnetic interactions are needed as well as intramolecular interactions [26,27]. Rajca and coworkers synthesized many two-dimensional polyradicals, and characterized them as superparamagnetic materials [28-30]. Nowadays, spins more than 5000 have been observed by using calixarene skeletons [29]. Theoretically and classically, the most typical two-dimensional system is so-called Mataga polymer *16* [31], as shown in Figure 7. This contains 1,3,5-phenylene unit as a ferromagnetic coupler. The Wannier functions span common atoms along two directions. The exchange integrals become positive, and significant ferromagnetic interactions are expected.

There are few studies on three-dimensional organic ferromagnets. However, in principle, π-stacked polyradicals are regarded as non-Kekulé systems, as noted above. For example, as shown in Figure 8, we can construct a possible three-dimensional organic ferromagnet *17* by cumulating the Mataga polymers in -ABABAB- fashion so that starred and unstarred atoms are linked. Then, considering perturbational molecular orbital methods, there should be also non-bonding crystal orbitals that are degenerate under the Hückel-like approximation. Though the analytical Wannier functions are too complex to calculate, they should be non-disjoint due to the starred-unstarred linkage. Then, Wannier functions span common atoms along the three directions, as schematically shown by spheres in Figure 8. There are few experimental proofs for three-dimensional organic ferromagnets at present. However, heterocyclic thia/selenazyl radicals synthesized by Oakley et al. are well-designed three-dimensional organic ferromagnets, of which magnetisms are realized through slipped π-stacking arrays [32].

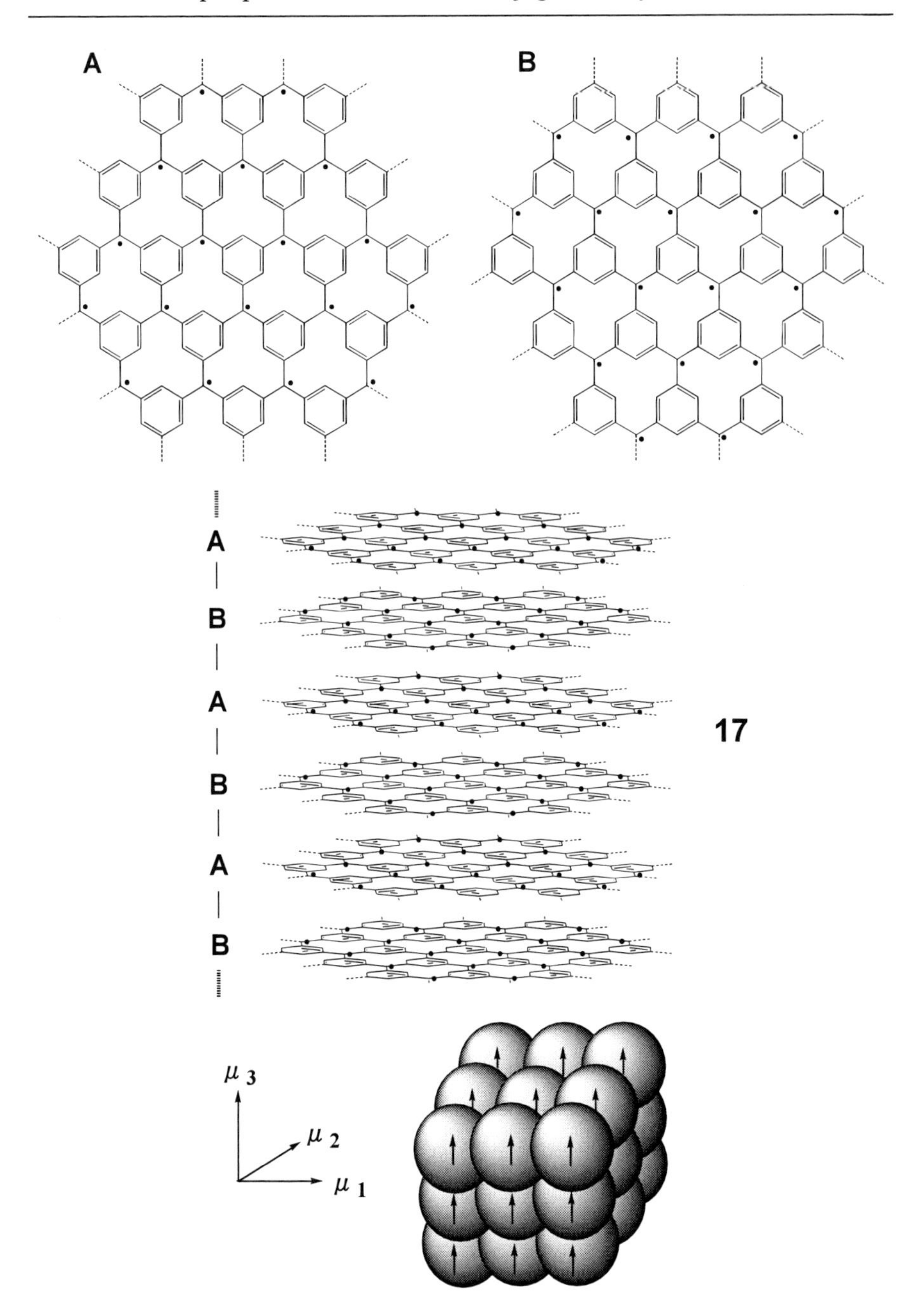

Figure 8. Possible three-dimensional organic ferromagnet *17* (excerpted from reference [19]).

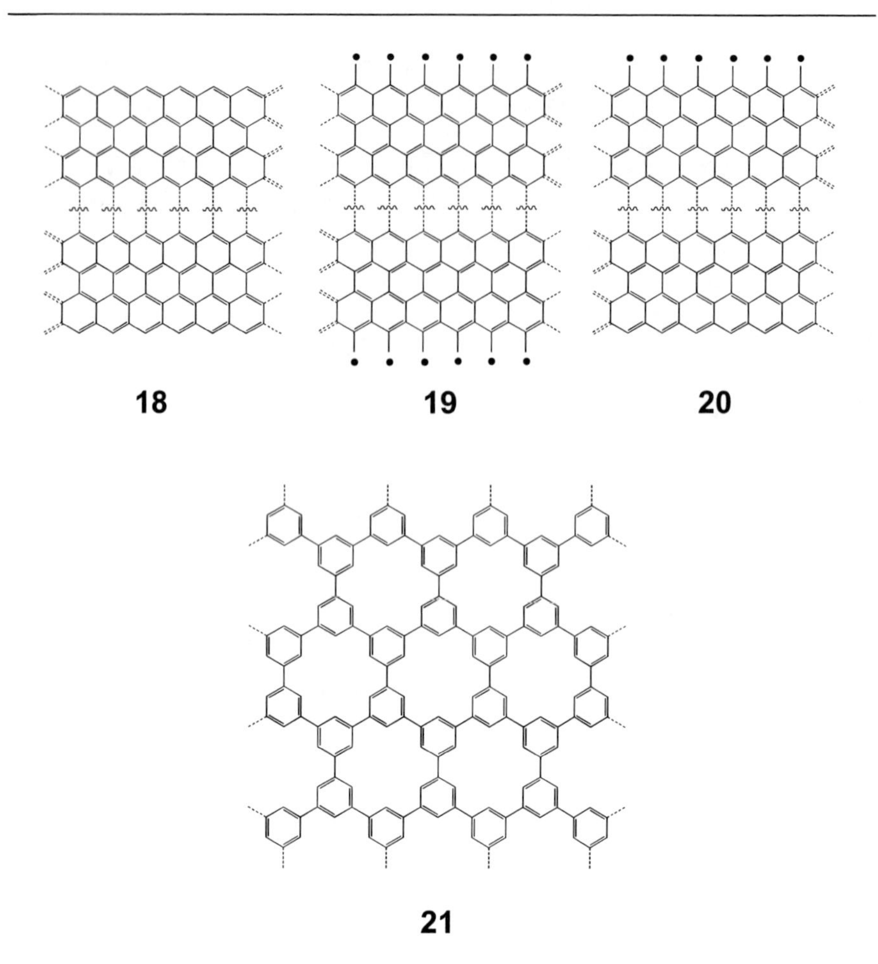

Figure 9. Graphene ribbons with zigzag edges *18*, two-sided Klein edges *19*, one-sided Klein edges *20*, and porous graphene *21*.

6. Other Applications

Finally, some applications toward nano-scale material science are referred. Since graphene is discovered [33,34], nano-scale carbons have attracted many chemists. Studies on graphenes have been done mainly toward conductive materials taking advantage of the tunable electronic states. On the other hand, magnetic properties on graphene derivatives have not been well studied due to the large band width and the Dirac point. It is well known that graphene nano ribbons with zigzag edges *18* (Figure 9) have semi-flat bands at $|k|>\pi 2/3$ [35].

If spin cancelations do not occurs, the partial flat bands themselves are interesting as possible necessary conditions for ferromagnetism. Klein suggested graphenes with two-sided methylene-edges *19*. They have semi-flat bands at $|k|<\pi 2/3$ [36]. The partial flat bands are also interesting, but the electronic states are probably 'disjoint', judging from the molecular symmetry. Graphenes with one-sided Klein edges *20* are probably ferromagnetic, because they can be regarded as non-Kekulé polymers with fully flat non-bonding bands [37]. Nowadays, ferromagnetism in defective HOPGs (highly oriented pyrolytic graphites) has often been observed [38,39]. This can be analyzed by flatness of frontier bands resulted from grain boundaries with zigzag edges [40]. In general, edge states and/or defective states are unstable, and the synthesis is also difficult. However, discovery of porous graphenes encouraged us toward flat-band ferromagnetism based on purely organic polymers [41,42]. Porous graphenes *21* were synthesized by aryl-aryl coupling reactions on metal surfaces such as copper, silver, and gold [41,42]. Immediately, we found that porous graphenes and porous nano ribbons have flat bands at all the wavenumber region [43]. It is important that porous graphenes are obtained as nearly planar two-dimensional polymers. Then, the electronic states are well defined, and the flat bands appear due to nodal character of the phenylene units. We note that the flat bands are *not* non-bonding. In neutral states, they should be non magnetic. However, when they are oxidized or reduced properly, ferromagnetic interactions are expected due to the flatness of the frontier bands.

These materials have flat bands at the frontier levels. The possible flat-band ferromagnetism will be described by effective exchange integrals between the Wannier functions. Band structures of these materials have been reported in several references [37,40,43]. Though calculations of their Wannier functions are considerably complex, some attempts have been done by the author for one-sided Klein-edge graphenes [37] and porous graphenes [44]. To calculate Wannier states, numerical integrations should be programed systematically. Under Hückel-level approximations, analytical solutions are often available. However, in general, Wannier functions at semi-empirical or *ab initio* levels need quite complex numerical integrals. Programs for maximally localized Wannier functions developed by several workers [45] are probably useful for complex systems. Essentially, amplitude patterns of Wannier functions should be not sensitive to the approximation levels or basis sets. Therefore, not only Hückel levels but also semi-empirical, *ab initio*, and DFT wavefunctions are available for the Wannier analysis.

CONCLUSION

Advances in spin-preference rules for polyradicals were reviewed. Through studies on exceptional species that violate the classical molecular orbital (MO) and/or valence bond (VB) theories, localized orbital methods have been found to solve the controversial situations on biradicals and polyradicals. In particular, Wannier functions as maximally localized orbitals are powerful principles to design for organic ferromagnets. Whether or not the Wannier functions span common atoms determines the spin preference of the system. When the adjacent Wannier functions span common atoms, the system is ferromagnetic. On the other hand, when Wannier functions span no common atoms, the system is antiferromagnetic. Nowadays, localized orbital methods are better than the conventional two rules in that some exceptional cases that violate classical predictions are consistently described. *Ab initio* calculations including density-functional theory (DFT) also supported the validity of the localized orbital methods.

ACKNOWLEDGMENTS

Thanks are due to Springer and Elsevier B. V. for permission to use figures in the original papers [19,21].

REFERENCES

[1] H. C. Longuet-Higgins, *J. Chem. Phys.* 18 (1950) 265.
[2] P. Dowd, *Acc. Chem. Res.* 5 (1972) 242.
[3] P. G. Wenthold, J. Hu, R. R. Squires, W. C. Lineberger, *J. Am. Chem. Soc.* 118 (1996) 475.
[4] S. Kato, K. Morokuma, D. Feller, E. R. Davidson, W. T. Borden, *J. Am. Chem. Soc.* 105 (1983) 1791.
[5] P. Du, W. T. Borden, *J. Am. Chem. Soc.* 109 (1987) 930.
[6] D.A. Hrovat, W.T. Borden, *J. Mol. Struct. (THEOCHEM)* 398–399 (1997) 211.
[7] T. Matsumoto, N. Koga, H. Iwamura, *J. Am. Chem. Soc.* 114 (1992) 5448.

[8] T. Matsumoto, T. Ishida, N. Koga, H. Iwamura, *J. Am. Chem. Soc.* 114 (1992) 9952.
[9] D. A. Hrovat, W. T. Borden, *J. Am. Chem. Soc.* 116 (1994) 6327.
[10] A. A. Ovchinnikov, *Theor. Chim. Acta* 47 (1978) 297.
[11] W. T. Borden, E. R. Davidson, *J. Am. Chem. Soc.* 99 (1977) 4587.
[12] W. T. Borden, *Mol. Cryst. Liq. Cryst.* 232 (1993) 195.
[13] W. T. Borden, H. Iwamura, J. A. Berson, *Acc. Chem. Res.* 27 (1994) 109.
[14] Y. Aoki, A. Imamura, *Int. J. Quantum Chem.* 74 (1999) 491.
[15] M. Hatanaka, R. Shiba, *Bull. Chem. Soc. Jpn.* 82 (2009) 206.
[16] A. Izuoka, S. Murata, T. Sugawara, H. Iwamura, *J. Am. Chem. Soc.* 107 (1985) 1786.
[17] A. Izuoka, S. Murata, T. Sugawara, H. Iwamura, *J. Am. Chem. Soc.* 109 (1987) 2631.
[18] H. M. McConnell, *J. Chem. Phys.* 39 (1963) 1910.
[19] M. Hatanaka, *Theor. Chem. Acc.* 129 (2011) 151.
[20] N. Marzari, D. Vanderbilt, *Phys. Rev. B* 56 (1997) 12847.
[21] M. Hatanaka, *Chem. Phys.* 392 (2012) 90.
[22] M. Hatanaka, R. Shiba, *Bull. Chem. Soc. Jpn.* 81 (2008) 966.
[23] N. Tyutyulkov, P. Schuster, O. Polansky, *Theor. Chim. Acta* 63 (1983) 291.
[24] N. N. Tyutyulkov, S. C. Karabunarliev, *Int. J. Quantum. Chem.* 29 (1986) 1325.
[25] M. Mitani, Y. Takano, Y. Yoshioka, K. Yamaguchi, *J. Chem. Phys.* 111 (1999) 1309.
[26] H. Iwamura, *Adv. Phys. Org. Chem.* 26 (1990) 179.
[27] H. Iwamura, *Proc. Japan Acad. Ser. B* 81 (2005) 233.
[28] A. Rajca, *Chem. Rev.* 94 (1994) 871.
[29] A. Rajca, J, Wongsriratanakul, S. Rajca, *Science* 294 (2001) 1503.
[30] A. Rajca, *Adv. Phys. Org. Chem.* 40 (2005) 153.
[31] N. Mataga, *Theor. Chim. Acta* 10 (1968) 372.
[32] C. M. Robertson, A. A. Leitch, K. Cvrkalj, R. W. Reed, D. J. T. Myles, P. A. Dube, R. T. Oakley, *J. Am. Chem. Soc.* 130 (2008) 8414.
[33] K. S. Novoselov, A. K. Geim, S. V. Morozov, D. Jiang, Y. Zhang, S. V. Dubonos, I. V. Grigorieva, A. A. Firsov, *Science*, 306 (2004) 666.
[34] K. S. Novoselov, A. K. Geim, S. V. Morozov, D. Jiang, M. I. Katsnelson, I. V. Grigorieva, S. V. Dubonos, A. A. Firsov, *Nature*, 438 (2005) 197.

[35] M. Fujita, K. Wakabayashi, K. Nakada, K. Kusakabe, *J. Phys. Soc.* Jpn. 65 (1996) 1920.
[36] D. J. Klein, *Chem. Phys. Lett.* 217 (1994) 261.
[37] M. Hatanaka, *Chem. Phys. Lett.* 484 (2010) 276.
[38] J. Červenka, M. I. Katsnelson, C. F. J. Flipse, *Nature Physics*, 5 (2009) 840.
[39] P. Esquinazi, R. Höhne, *J. Mag. Mag. Mater.* 290-291 (2005) 20.
[40] M. Hatanaka, *J. Mag. Mag. Mater.* 323 (2011) 539.
[41] M. Bieri, M. Treier, J. Cai, K. Aït-Mansour, P. Ruffieux, O. Gröning, P. Gröning, M. Kastler, R. Rieger, X. Feng, K. Müllen, R. Fasel, *Chem. Commun.* (2009) 6919.
[42] M. Bieri, M.-T. Nguyen, O. Gröning, J. Cai, M. Treier, K. Aït-Mansour, P. Ruffieux, C. A. Pignedoli, D. Passerone, M. Kaster, K. Müllen, R. Fasel, *J. Am. Chem. Soc.* 132 (2010) 16669.
[43] M. Hatanaka, *Chem. Phys. Lett.* 488 (2010) 187.
[44] M. Hatanaka, *J. Phys. Chem. C* 116 (2012) 20109.
[45] A. A. Mostofi, J. R. Yates, Y.-S. Lee, I. Souza, D. Vanderbilt, N. Marzari, *Comput. Phys. Commun.* 178 (2008) 685.

INDEX

A

absorption spectra, 71, 72
access, 49
acne vulgaris, 6
acupuncture, viii, 2, 6, 13, 16, 17, 18, 23, 24
adhesion, 39
adjustment, vii, 1, 8, 12, 18
advancement, 34
AFM, 90, 91
age, vii, 2, 12
aluminium, 42
amplitude, 80, 85, 95
analgesic, 5
anisotropy, viii, 33, 35, 36, 43, 44, 45, 46, 47, 49, 51
annihilation, 49
apoptosis, 29
arterial hypertension, 17
arthralgia, 6
atomic orbitals, 84
atoms, 58, 66, 67, 78, 80, 86, 91, 92, 96
atopic dermatitis, 6, 29
attachment, 14

B

base, 5, 24, 34
batteries, 13
Belarus, 1, 23, 24, 28, 30, 31, 32
benefits, 42
biological activity, 30, 31, 32
biological systems, 25
biosynthesis, 31
blood, viii, 2, 3, 5, 6, 13, 18, 23, 24
bonding, 78, 82, 84, 88, 92, 95
bone, 6, 19, 27, 32
bone marrow transplant, 27

C

cadmium, 3
carbon, 78, 84
carpal tunnel syndrome, 28
case study, 46
catheter, 18
cation, 66, 70, 71, 74
CD95, 29
cell line, 31
ceramic, 34
certification, 5
challenges, viii, 33, 36
chemical, 5, 74
chemotherapy, 26

children, 29
China, 4
cholesterol, 6
circulation, 5
CIS, 4
cleaning, 17
CO2, 60
coherence, 30
colitis, 6
combined effect, 5
commercial, 13
competition, 35
complications, 29
composition, 35
compounds, 29, 58, 59, 60, 64, 76
computer, 34
conductivity, 51
conductor, 34
configuration, 43, 44, 45, 47, 49, 56, 57, 59, 89
confinement, 45
construction, ix, 14, 24, 78
consumption, 13
contour, 45
controversial, 80, 96
cooling, 38
coordination, viii, 55, 57, 75
copolymer, 42, 43
copper, 3, 95
copyright, 38, 41, 44, 48
coronary heart disease, 6
correlation, 64
correlations, 89
cost, 13, 36, 37, 42, 52
critical analysis, 28
crystal structure, 58
crystalline, 43, 44, 45, 46
crystals, 19
cure, 42
curing process, 42

D

decay, 85
defects, 43, 92
deficiency, 31
deformation, 52
degenerate, 78, 80, 82, 85, 89, 91, 92
depolarization, 13
deposition, 38, 45
depth, 19, 23
derivatives, 80, 81, 94
dermatitis, 6
dermatology, 6
deviation, 52
DFT, ix, 78, 86, 89, 91, 95, 96
diabetes, 3
diode laser, 27
diodes, viii, 2, 13, 14, 19, 21, 24
direct action, 3
discomfort, 14
diseases, 3, 4, 6
disposable intravenous tips, viii, 2
distribution, vii, 1, 8, 9, 10, 11, 12, 48, 49, 51
divergence, 13, 14, 15, 17, 24
domain structure, 44
dopants, 91
double bonds, 78
drug consumption, 24
drug therapy, 3
duodenal ulcer, 18
durability, 13

E

eczema, 6
edema, 5
electric field, 7
electron, 43, 56, 57, 59, 60, 61, 62, 64, 68, 85
electron beam lithography, 43
electronic structure, 57

electrons, 60, 78, 80
electroplating, 37, 38, 40
emitters, 13
endocrine, 5
energy, 5, 26, 27, 35, 43, 44, 45, 46, 57, 70, 89, 90, 91
engineering, 3
entropy, 69, 70, 71
epitaxial films, 44
equilibrium, 46
equipment, vii, 2, 14
ESR, 79, 90
etching, 38, 40, 42
Europe, 5
evidence, 4
excitation, 35, 66
exploitation, viii, 33
exposure, 6, 7, 12, 13, 18, 21, 23, 24, 31

F

fabrication, 37, 51, 52, 57
FDA, 4
ferrite, 8
ferromagnetic, 34, 59, 60, 65, 69, 70, 73, 86, 88, 91, 92, 95, 96
ferromagnetism, 70, 73, 95
ferromagnets, 59, 91, 92, 96
fiber, 3, 13, 24
films, 42
first generation, 13
fish, 17, 30, 31
flatness, 91, 95
flexibility, 52
Food and Drug Administration, 4
force, 69
formation, vii, 1, 5, 12, 42, 48
foundations, 75
fractures, 6
free energy, 66
freedom, 80, 84
frostbite, 6

G

gastric ulcer, 18
gastritis, 6
germanium, 40
Gibbs energy, 70, 71
gingivitis, 6
glass transition, 38
grain boundaries, 95
grain size, 51
Great Britain, 27
grounding, 13
growth rate, 34

H

Hamiltonian, 64
hardware implementation, vii, 1
Hartree-Fock, 89, 91
head and neck cancer, 26
healing, 27
heart disease, 17
heat capacity, 68
height, 8
helium, 2, 27
hepatitis, 6
herpes, 6, 29
high recording density, viii, 33, 34
human, 25, 31
human body, 13
humidity, 56
Hungary, 4
hybrid, vii, viii, 55, 57, 62, 66, 70, 75
hydrocarbons, 78
hydrogen, 43
hydroxide, 58
hypertension, 6, 29
hysteresis, 48, 68
hysteresis loop, 46, 47, 49

I

image(s), 41, 44, 48, 49, 79
immune system, 19
immunotherapy, 5
imprinting, viii, 34, 38, 39, 42
induction, vii, 1, 6, 8, 9, 10, 11, 12, 17, 20, 21
industry, viii, 33, 51
infection, 29
infertility, 6, 30
inflammation, 27
inflammatory disease, 6
injury, 3, 19, 32
internal laser therapy, vii, 2
intratissue therapy, viii, 2, 13
intravascular laser therapy, viii, 2
intravenously, 18
ions, 58, 59, 60
Ireland, 33
irradiation, viii, 6, 13, 29, 30, 31, 55, 56, 57, 64, 65, 70, 71, 72, 73
islands, 36
isomerization, 62, 65, 70, 72, 74
Italy, 4

J

Japan, 55, 77, 97
joints, 6, 19, 32

K

Kazakhstan, 25
KBr, 71, 72

L

laptop, 34
larva, 31
laser radiation, 2, 3, 4, 5, 6, 7, 12, 13, 15, 17, 19, 20, 21, 25, 26, 29, 30, 31, 32
laser-induced acupuncture, viii, 2
lasers, 2, 13, 14, 19, 20, 21, 26, 27
layered double hydroxides, 58
lead, 23, 86
LED, 12, 20, 21
lens, 14, 16, 18
lesions, 13
lichen planus, 6
ligand, 57, 59
light, viii, 6, 7, 13, 14, 17, 19, 24, 31, 55, 56, 57, 64, 65, 72, 74
light beam, 24
light scattering, 7
lipid peroxidation, 6
lithography, vii, 37, 42, 43, 51, 52
liver, 19
localization, 7, 13, 86
lymphocytes, 25
lymphoid, 31

M

macromolecules, 17
magnet, vii, 2, 7, 8, 9, 10, 11, 12, 15, 16, 17, 44, 45, 64
magnetic field, vii, 1, 5, 6, 7, 8, 9, 12, 15, 17, 19, 30, 32, 36, 45, 49, 51, 68
magnetic force microscope, 47
magnetic induction, vii, 1, 6, 8, 9, 10, 11, 12, 20
magnetic materials, vii, viii, 1, 12, 33, 55
magnetic moment, viii, 33
magnetic particles, 34, 49
magnetic properties, viii, 33, 43, 58, 59, 61, 75, 94
magnetic structure, 39, 43, 44, 45
magnetism, viii, 43, 55, 56, 57, 59, 64, 75, 92
magnetization, viii, 34, 35, 36, 43, 44, 45, 46, 47, 48, 49, 50, 68, 72, 73
magnetoresistance, 34
magnets, viii, 34, 43, 44, 45, 55, 64, 76, 92

magnitude, 89
majority, 3
management, 27
mass, 13, 51
mast cells, 5
mastitis, 6
materials, vii, viii, 35, 36, 39, 42, 44, 55, 58, 70, 92, 94, 95
matrix, 41
measurement, 68
measurements, 79
media, viii, 33, 34, 36, 42, 43, 51, 52
medical, vii, 1, 2, 5, 6, 7, 12
medicine, 2, 4, 24, 27, 28, 31
membranes, 17
memory, 51
MEMS, 53
meta-analysis, 27
metabolism, 7
metal ions, 66
methyl groups, 59
microcirculation, 5
micrometer, 44
microscope, 14, 47
microstructure, 49
mineral water, 29
modelling, viii, 33
models, 86, 88, 89
molecular orbital (MO), ix, 77, 78, 96
molecules, 58, 66, 75, 80, 81
Moscow, 25, 26, 28, 30, 31, 32
moulding, 37
myalgia, 6
myelomeningocele, 26
myocardial infarction, 17

N

nanofabrication, 51
nanoimprint, 37, 38, 39, 42, 43, 51, 52
nanolithography, 53
nanometer, 40
nanometer scale, 43
nanoparticles, viii, 33, 49
nanopatterning techniques, viii, 33
nanostructures, vii, v, 33
National Academy of Sciences, 1
NCS, 70
Nd, 3, 19, 32
neon, 2, 27
neuralgia, 6
neuropathy, 6
neutral, 95
neutrophils, 29
nitrogen, 3, 59
nitroxide, 80, 90
nucleation, 49
nucleic acid, 5

O

optical fiber, 14, 24
oral cavity, 18
orbit, 59
organic polymers, 95
organism, 2
organs, 5, 6, 13
osteoarthritis, 28
ox, 67, 68
oxygen, 31, 58

P

pain, 3, 27
pairing, 57
parallel, viii, 33, 44, 50, 78, 80
pathology, vii, 2, 12
peptic ulcer, 6
perinatal, 3
periodicity, 36
periodontitis, 6
permeability, 5, 34
permission, 38, 41, 44, 48, 96
PES, 90
pharmaceutical, 3, 4

phase diagram, 70, 71
phase transitions, 65, 73
phosphates, 5
photoblockage intrajoint, vii, 2, 13
photomagnetism, viii, 55, 56, 57, 62, 74, 75
physical therapy, 2, 28
physics, viii, 45, 55
pilot study, 26
pitch, 52
placebo, 4
pleasure, 75
PM3, 90
pneumonia, 29
polarity, 9
polarization, ix, 14, 30, 31, 77, 78, 79, 80, 88, 90
polydimethylsiloxane, 41, 43
polymer(s), viii, 14, 18, 38, 39, 40, 42, 55, 57, 75, 83, 88, 92, 95
polystyrene, 41, 43
prevention, 26, 27, 29
primary prophylaxis, 29
principles, vii, 8, 14, 24, 96
proliferation, 19
propagation, 16
prophylaxis, 2
proteins, 5
prototypes, 86
psoriasis, 6
pyrolytic graphite, 95

Q

quantum-chemical calculations, 86, 91
quartz, 14
questionnaire, 27

R

radiation, v, vii, 1, 2, 3, 4, 5, 6, 7, 8, 12, 13, 14, 15, 16, 17, 18, 19, 20, 21, 23, 24, 25, 26, 29, 30, 31, 32
radical pairs, 89
radicals, viii, 77, 80, 81, 92
radiotherapy, 26
radius, 8
reactions, 5, 95
reality, 4
receptors, 29
rehabilitation, 2
reliability, 13
repair, 26, 27
repulsion, 80, 85
resolution, 37, 43, 52
resonator, 18
response, 27, 73
rheology, 5
rhodopsin, 74
RIE, 38
ring magnet, vii, 1, 7, 8, 9, 11, 12, 45
risk, 35
room temperature, 35, 38, 42, 65, 66, 72, 73, 75
rules, vii, viii, 77, 78, 80, 88, 90, 91, 96
Russian literature, 3

S

saturation, 35, 47, 49
scaling, 34
science, 94
scientific understanding, 4
scope, 14, 35
secretion, 29
seed, 38
self-assembly, 52
semiconductor, 40, 42
semiconductor lasers, 13, 18
sensitivity, 7, 17

sham, 28
shape, 45, 46, 47, 49
shortage, 42
showing, 37
side effects, 3, 7, 24
signal-to-noise ratio, viii, 33
signs, 82, 91
silver, 95
simulation, 4
Singapore, 26
SiO2, 36
skin, 7, 13
skin diseases, 3, 6, 19
soft magnetic materials, vii, 1, 12
solution, 7, 14, 42, 46, 51
specialization, 2
species, 58, 88, 96
spin, vii, viii, 34, 38, 42, 45, 56, 57, 59, 64, 66, 69, 70, 71, 74, 77, 78, 79, 80, 81, 84, 86, 88, 89, 90, 91, 95, 96
spindle, 34
spine, 6
spring, 16, 17
stability, 35, 36, 82
stabilization, 50
state(s), ix, 18, 19, 44, 45, 49, 50, 56, 57, 59, 66, 75, 77, 78, 79, 80, 81, 84, 86, 88, 89, 90, 91, 94, 95
steel, 10, 14
sterile, 13, 18, 21, 24
stimulation, 5
stomatitis, 6
storage, viii, 33, 34, 43, 45, 49, 51
storage media, viii, 33
stress, 74
strontium, 8
structure, 19, 38, 39, 41, 44, 45, 58, 66
substrate(s), 37, 38, 39, 41, 43, 52
sulfate, 29
superparamagnetic, 35, 51, 92
susceptibility, 63, 64, 65, 68, 79
symmetry, 7, 51, 80, 89, 90, 95
symptoms, 26
synthesis, 5, 6, 19, 92, 95

T

target, 51
technical assistance, 25
techniques, viii, 4, 6, 33, 36, 47
technology(ies), viii, 12, 17, 28, 33, 34, 35, 36, 37, 42, 51
temperature, 21, 35, 36, 38, 51, 56, 57, 60, 61, 65, 68, 69, 70, 72, 75, 89
temperature dependence, 59
testing, 4
therapeutic effects, 5, 7, 13
therapy, vii, 1, 2, 3, 5, 6, 7, 8, 12, 14, 17, 18, 19, 21, 23, 24, 25, 26, 27, 28, 29, 30, 32
thermal energy, 35
thermal evaporation, 38
thermal stability, 35
tissue, 14, 27
total energy, 43, 44, 46
tracks, 34
transcutaneous laser-induced blood therapy, viii, 2
transformation, 80, 82, 84, 85
transition metal, 56
transition temperature, 70
transparency, 19
transport, 19
trauma, 6
treatment, 2, 3, 4, 6, 14, 26, 28, 29, 30, 39
trial, 26, 27, 28
triggers, 74
tuberculosis, 26

U

uniform, 43
USA, 4
USSR, 4, 25
UV, 52

UV irradiation, 65, 71, 72, 73
UV light, 38, 42, 65, 70, 71, 73
UV radiation, 3

V

valence, ix, 57, 66, 77, 78, 90, 96
valence bond (VB), ix, 77, 78, 96
vapor, 3
vascular diseases, 6, 32
versatility, 75
vessels, 6, 24
vibration, 69
Vietnam, 30
viscosity, 42

W

Wannier functions, ix, 77, 83, 85, 86, 87, 88, 91, 92, 95, 96
water, 59
wavelengths, 18
wear, 39
workers, 29, 95
worldwide, 19
wound healing, 25, 27, 28

X

X-ray diffraction(XRD), 62, 64

Y

yttrium, 19

Z

zonal external magnetolaser therapy, vii, 2